Landscape-Level Long-Term Biological Research and Monitoring Plan for the Crane Trust

Andrew J. Caven[1,†], Joshua D. Wiese[1], Bethany L. Ostrom[1,*], Kelsey C. King[1,§], Jenna M. Malzahn[1,‡], David M. Baasch[1], and Brice Krohn[1]

[1]Platte River Whooping Crane Maintenance Trust, Wood River, NE, USA

Version 2.0 – 31 October 2022

*Contact: bostrom@cranetrust.org; †Current affiliation: International Crane Foundation, Baraboo, WI; §Current affiliation: Washington State University–Vancouver, Vancouver, WA; ‡Current affiliation: Auburn University, Auburn, Alabama.

ISBN 978-1-60962-262-6 paperback

ISBN 978-1-60962-263-3 ebook

DOI:10.32873/unl.dc.zea.1332

Zea Books are published by the University of Nebraska–Lincoln Libraries.

Electronic (pdf) edition available online at
https://digitalcommons.unl.edu/zeabook/

UNL does not discriminate based upon any protected status.
Please go to http://www.unl.edu/equity/notice-nondiscrimination

Nebraska UNIVERSITY OF
Lincoln

Table of Contents

Background and Purpose

Our obligation is to make sure we are effectively utilizing science to meet the objectives of the Platte River Whooping Crane Maintenance Trust (1981) laid out in its charter "to rehabilitate and preserve a portion of the habitat for Whooping Cranes and other migratory birds in the Big Bend reach of the Platte River between Overton and Chapman (i.e., Central Platte River Valley), Nebraska". The original declaration is aimed at maintaining "the physical, hydrological, and biological integrity of the Big Bend area as a life-support system for the Whooping Crane and other migratory species that utilize it." It was clear from the institution's founding that to accomplish this goal it was necessary to study the effectiveness of land conservation and management actions in providing habitat for Whooping Cranes and other migratory bird species. Quality habitat necessarily comprises all the components that Whooping Cranes and other migratory bird life require to complete their migrations –food and shelter– including nutrient rich diet items such as invertebrates, vascular plants, herpetofauna, fish, and small mammals as well as suitable roosting and foraging locations including wide braided rivers and undisturbed wet meadows (Allen 1952; Steenhof et al. 1988; Geluso 2013; Caven et al. 2019, 2021). Article "A" of the Crane Trust's (1981) declaration is "to establish a written habitat monitoring plan which can be used to describe change in…[habitat] within the Big Bend of the Platte River…utilized by Sandhill Cranes and Whooping Cranes…." Following initial inventories including avian (Hay and Lingle 1982), vegetation (Kolstad 1981; Nagel 1981), small mammals (Springer 1981), herpetofauna (Jones et al. 1981), insects (Ratcliffe 1981), and fish (Cochar and Jenson 1981), a variety of excellent research has continued at the Crane Trust (https://cranetrust.org/conservation-research/publications/). However, despite the clarity of the Trust's original declaration, long-term habitat monitoring has not progressed unabated throughout the history of the Crane Trust.

Long-term monitoring has the ability to inform a natural resources program about landscape-level biological changes taking place in space and over time. This is very important in the era of global climate change. Changes in phenology, such as plant species fruiting and flowering timing, breeding and migratory bird arrivals in the Central Platte River Valley, and small mammal and herpetological breeding activity are very important signifiers of ecological change (Bradley et al. 1999). These cannot be assessed in a year-to-year manner, as the weather varies widely across individual years, however, can be assessed at the decadal timescale and beyond. The impacts of management actions such as controlled burning or the removal of trees and shrubs from the landscape also cannot be assessed on a short-term basis (Briggs et al. 2002; Collins and Calabrese 2012). Ecological phenomena mostly take place across several years, decades, lifetimes, and careers (Bragg and Hulbert 1976; Williams 1978). We need a monitoring system that can stay in place beyond the tenure of any one natural resource scientist and provide the Crane Trust with a better understanding of the dynamic changes taking place on the landscape, while giving our organization an opportunity to contribute to the scientific literature on the biggest conservation issues of our time, such as species adaptions to climate change (changing distributions and behaviors) and the impact of creative management practices in those contexts.

Landscape-level and temporal environmental changes can only be assessed accurately by examining long-term trends. The National Park Service, the Bureau of Land Management, the Forest Service, and the Fish and Wildlife Service now all have inventory and monitoring programs at various stages of development, mostly operating at regional and district-level scales (Toevs et al. 2011; USFWS 2013; USNPS 2015; Powell 2000). These federal agencies have gone to great lengths to standardize their monitoring programs across similar ecoregions, allowing comparisons across both time and space. The Long-Term Ecological Research Network (LTER) funded by the National Science Foundation has dedicated the last 35 years to long-term ecological research oriented toward studying phenomena that take place across decades and at continental and global scales (LTER 2018). The National Science Foundation launched a new program in 2011 called the National Ecological Observatory Network (NEON), which focuses on standardizing biological monitoring efforts across the United States and its affiliated territories. They believe it will provide more consistency in ecological variable measurement across inconsistent landscapes to create more comparable data across North America and beyond (NEON 2020). Additionally, a high number of states including Iowa, Washington, Minnesota, Florida, California, and Texas have their own biological monitoring programs run by state natural resource agencies, often more loosely defined than federal programs, but functioning and collecting data across a diversity of state lands and across time (See MN DNR 2016 for an example).

It is not only federal and state agencies that are undertaking these efforts. Non-profit Organizations such as the Ladybird Johnson Wildflower Center in Austin, TX, and the Tall Timbers Research Station and Land Conservancy in the panhandle of Florida, have been expertly involved in inventory, monitoring, and long-term research efforts on their lands for decades (LBJWFC 2022, Tall Timbers 2015). Tall Timbers actually pioneered a cutting-edge fire program that looks at the differential impacts of fire frequency and seasonality on plant and wildlife communities over time. This is where I see us fitting in; utilizing our standardized biological monitoring program to answer long-term research questions related to the effects of land management, regional and local land use changes, and climate change on the ecological community, in particular the vegetative and avian communities. This allows us to participate in the effort to address the largest conservation concerns in our time and region through sound science and targeted land management.

Implementing a long-term monitoring program will help us more critically assess the impacts of our management actions to better understand their impacts upon Whooping Cranes, Sandhill Cranes, and the other bird species that pass through and breed in the Big Bend region of the Platte River in Nebraska. It is also our responsibility to publish, in scientific journals and management publications, the results of our research so that we may inform our colleagues in the conservation community and maintain a public record of our research program at the Crane Trust. We need to better understand dynamic ecological change in the long view by implementing an achievable, straight-forward research and monitoring strategy on the Crane Trust's properties. We have an opportunity to collect ecological information on an important mosaic of wet meadows, river channels, sloughs, woodlands, and tallgrass prairies, across a diversity of soil types and biotic communities, under a variety of management regimes. So much incredible research has been done at the Platte River Whooping Crane Maintenance Trust over the last 40 years. One way in which we can continue this legacy is via the continued implementation of our long-term research and monitoring efforts, allowing us to be cooperators in the great task of understanding the "long-scale" ecological changes taking place in the Great Plains, and in our world. This is the future trend of ecological science and land management; we are moving from shorter-term experiments to long-term data collection as a standard practice for understanding the ecological systems we inhabit.

Our Contexts

Increased water appropriation and damming starting in the early 1900s has irreparably altered the hydrology of the Platte River, impeding the once massive floods creating the wide, shallow, temporary flows followed by seasonal decreases that the Platte River is known for (Williams 1978). Increased settlement and agriculture created an increased number of fire breaks also disrupting the cyclical influence of fire, which aided flooding in setting back woody vegetation development (Bragg and Hulbert 1976). The grazing regimes of the Platte River Basin and Islands have been significantly altered for an even longer period of time, going back to the decimation of the Bison by the late 1800s. Reports of Bison on the Platte in substantial numbers were recorded as late as 1857 (Hart 2001). As Currier (1982) notes, disturbance regimes-flooding, fire, and grazing, would have maintained the Platte River's vegetated islands as a mosaic of early and mid-succession communities based upon various models. Currier (1982) suggested that no area of the central Platte has yet reached the potential non-disturbance climax vegetation community dominated by *Fraxinus, Ulmus* [*americana*], *Celtis, Elaeagnus, Morus,* and *Juniperus* genera woody species. *Populus* and *Salix* genera species need open mudflats to regenerate and only live to be about 100 years old; those species now have a harder time germinating at a significant scale as open mudflats have declined significantly without recurrent large-scale flooding disturbances. Currier (1982) hypothesized that the aforementioned woody species- *Fraxinus, Celtis,* etcetera, could take over after the now dominant *Populus* (Cottonwood) and *Salix* (Willow) species declined. However, one projected influence unforeseen by Currier (1982) was the Emerald Ash Borer (*Agrilus planipennis*), which could impact this successional process in the coming decades (Wiese and Caven 2018). Additionally, Caven and Wiese (2022) found that exotic-invasive Siberian Elm (*Ulmus pumila*), which was not detected in the early 1980s regionally, is now the most widely distributed tree in portions of the Big Bend reach. These forest community dynamics alone highlight the need for long-term research.

Boettcher and Johnson (2000) argue that several places on the river had very old forests, but early settlers cut them down, and by the time a great number of people descended on central Nebraska to farm, an open treeless river had been produced. They argue the subsequent afforestation began in 1930 and has continued until today. However, the wooded areas would likely have been away from the fluctuating river banks as a result of once massive floods (Williams 1978; Junk et al. 1989). Additionally, periodic wildfires would likely have eliminated all but the most fire tolerant tree species, such as mature Cottonwoods (Bragg and Hulbert 1976; Hengst and Dawson 1994). Historic maps and photographs from the middle 1800s through the 1930s demonstrate that the woodlands would have been mostly on raised islands surrounded by wide open river channels with many unvegetated banks, and that the river would have generally been without trees except for scattered cottonwoods (US Bureau of Reclamation 1938; Currier 1982; Currier and Davis 2000; Schroeder 2015; Figure 1). However, as sustained flows, periodic flood pulses, and sediment transportation have been dramatically reduced trees have increasingly begun growing on the riverbanks and within former river channels (Williams 1978; Figure 1).

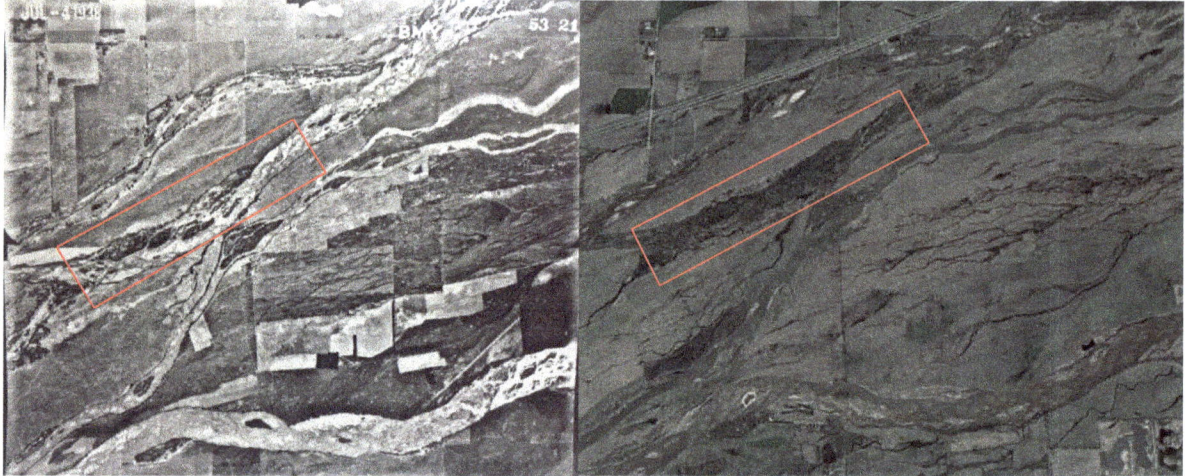

Figure 1. Photos of Mormon Island from 04 July 1938 and 21 April 2016 demonstrating tree encroachment along and within, especially the northern channels, of the central Platte River, Nebraska, USA (US Bureau of Reclamation 1938; Google Earth 2017). Area of particular interest outlined in red.

The Crane Trust property constitutes the most pristine tract of untilled land left in the Big Bend Region of the Platte River, a dynamic and important system, currently in the process of successional change. Relict tallgrass prairie has become a priceless resource throughout the central United States. It is estimated that over 97% of the tallgrass prairie in Nebraska has been lost (Noss et al. 1995). The once vast tallgrass prairies of eastern Nebraska are the most endangered ecosystem in the state with an estimated 99% of that system now eliminated (Ratcliffe and Hammond 2002). Areas of tallgrass prairie further west in the state remain somewhat more intact but depend on root access to subsurface moisture (subirrigated) provided by shallow groundwater (hyporheic hydrology) within braided prairie river valleys, and therefore are relatively isolated geographically (Kaul et al. 2006; Rolfsmeier and Steinauer 2010). Wet meadows generally exist along a hydrological gradient between lowland prairies and shallow marshes (Kantrud et al. 1989; Kaul et al. 2006; Rolfsmeier and Steinauer 2010; Tiner 2016). These systems experience periodic and temporary surface inundation, generally during the spring, but have saturated soils for longer durations of each year (Keddy 2010; Rolfsmeier and Steinauer 2010; Tiner 2016). Wet meadows in the Central Platte River Valley provide habitat for a range of species of concern, particularly waterbirds, such as the Whooping Crane (Bomberger Brown and Johnsgard 2013; Baasch et al. 2019).

The fertile soils of tallgrass prairies are ideal for agriculture and the depletion of contiguous habitat is attributed to the expansion of corn and soybean monocultures. As more virgin land was converted to cropland, populations of tallgrass-endemic species have sharply declined. It will be important to capitalize on quality past research by documenting the current state of the ecosystem for historic comparison then committing to consistent monitoring through standardized methods going forward to provide additional inferential power to understand trends taking place. This will help us to effectively focus our conservation and management efforts toward the areas of greatest concern to maintain and protect critical habitat for migratory Whooping Cranes, Sandhill Cranes, and other bird life.

Overview

The summer of 2015 saw the implementation of the long-term monitoring plan plot layout, vegetation monitoring, avian monitoring, and small mammal monitoring. Additionally in 2015, we piloted our butterfly species of concern monitoring program which became fully operational in 2017. We have also continued the ground water level monitoring (transducer) project started by Dr. Mary Harner, the former Director of Science at the Crane Trust. Additionally, we have continued the monitoring of slough fish species composition in both Calving Pasture Slough and Big Slough started by Greg Wright, former Wildlife Biologist at the Crane Trust. Finally, in 2018, we added an Anuran (Frogs and Toads) monitoring program involving call surveys. These are standard biological monitoring variables frequently used in monitoring plans because of their sensitivity to landscape-level changes, climate trends, and management actions (USNPS 2015; LTER 2018; Powell 2000). The National Park Service calls these monitoring variables, vital signs:

> "Vital signs monitoring tracks a subset of physical, chemical, and biological elements and processes…to represent the overall health or condition of park resources [and] known or hypothesized effects of stressors…. Monitoring results are used by park managers…to support management decision-making, park planning, research, [and] education [regarding] park resources" (USNPS 2015).

Monitoring variables are interrelated and vary together, but also can be viewed as outcome or dependent variables in more experimental contexts. For example, changes in vegetation often drive changes in avian and small mammal populations. For instance, woody plant encroachment on prairies denudes small mammal diversity and produces changes in the bird community present (Davis 2005; Horncastle et al. 2005). These biological community responses can be evaluated as dependent variables in response to management actions (independent variables) such as fire, woody species control, and grazing. A particular emphasis will be placed on the differential impacts of Bison verses Cattle grazing, fire timing and frequency, and the removal of riparian (Cottonwood) and upland (Eastern Red Cedar) tree species. All of the selected monitoring variables have been shown to be sensitive to changes in management regimes and environmental stressors such as drought. For instance, Bison grazing tends to produce more forb rich prairies than Cattle grazing, but this finding has not been widely validated (Steuter and Hildinger 1999; Rosas et al 2008; Caven et al. 2019). In this analysis, vegetation is the dependent variable, and grazing regime –Bison or Cattle– the independent variable. Another illustration of the phenomena we plan to study over time is highlighted by the research of Horncastle et al. (2005), which found that as Eastern Red Cedar encroachment increases, the diversity and species richness of small mammal populations decreases. In this case the diversity and species richness of the small mammal populations serves as the dependent variable, and the differential management of woody species serves as the independent variable. We will also be able to consider multiple variables as in Powell's (2006) study analyzing the combined impacts of Bison grazing and fire frequency on bird communities. Having a planned monitoring system in place will help us clearly understand similar processes over long periods of time and will help us to continue to build on the great biological work already completed at the Crane Trust.

Report Design

The various monitoring/research plans, including the plot layout, vegetation, avian community, small mammal, water level, slough fish, butterfly species of concern, anuran, aerial Sandhill Crane survey, Greater Prairie-chicken lek survey, Whooping Crane behavior, and Western Prairie Fringed Orchid monitoring plans are included below as chapters. This report is treated almost as an edited volume with consulted and in-text citations placed at the end of every protocol or chapter. After we address protocols regarding ongoing monitoring, we will discuss potential additional future research, which given available resources and/or collaborators, could make a significant contribution to our core monitoring program and our understanding of Platte River ecology, such as aquatic macroinvertebrate and river channel morphology research. A concluding remarks section will be placed at the end of this document.

References

Allen, R.P., and J.A. Livingston. 1952. The whooping crane. National Audubon Society, New York, New York, USA, 246 pp.

Baasch, D.M., P.D. Farrell, A.T. Pearse, D.A. Brandt, A.J. Caven, M.J. Harner, G.D. Wright and K.L. Metzger. 2019. Diurnal habitat selection of migrating Whooping Crane in the Great Plains. Avian Conservation and Ecology 14(1):6.

Bradley, N.L., A.C. Leopold, J. Ross, and W. Huffake. 1999. Phenological changes reflect climate change in Wisconsin. Proceedings of the National Academy of Sciences 96:9701-9704.

Bragg, T.B., L.C. Hulbert. 1976. Woody plant invasion of unburned Kansas bluestem prairie. Journal of Range Management 29(1):19-24.

Briggs, J.M., G.A. Hoch, and L.C. Johnson. 2002. Assessing the rate, mechanisms, and consequences of the conversion of tallgrass prairie to *Juniperus virginiana* forest. Ecosystems 5(6):578-586.

Bomberger Brown, M., and P.A. Johnsgard. 2013. Birds of the Central Platte River Valley and Adjacent Counties. Zea E-Books (No. 15), University of Nebraska-Lincoln, Lincoln, Nebraska, USA, 184 pp. <http://digitalcommons.unl.edu/zeabook/15>

Caven, A.J., E.M. Brinley Buckley, K.C. King, J.D. Wiese, D.M. Baasch, G.D. Wright, M.J. Harner, A.T. Pearse, M. Rabbe, D.M. Varner, B. Krohn, N. Arcilla, K.D. Schroeder, and K.F. Dinan. 2019. Temporospatial shifts in Sandhill Crane staging in the Central Platte River Valley in response to climatic variation and habitat change. Monographs of the Western North American Naturalist 11:33–76.

Caven, A.J., K.D. Koupal, D.M. Baasch, E.M. Brinley Buckley, J. Malzahn, M.L. Forsberg, and M. Lundgren. 2021. Whooping Crane (*Grus americana*) family consumes a diversity of aquatic vertebrates during fall migration stopover at the Platte River, Nebraska. Western North American Naturalist 81(4):592-607.

Caven, A.J., and J.D. Wiese. 2022. Reinventory of the Vascular Plants of Mormon Island Crane Meadows after Forty Years of Restoration, Invasion, and Climate Change. Heliyon 8(6):e09640.

Caven, A.J., J.D. Wiese, A. Fowler, D.H. Ranglack. 2019. Vegetation community composition within Bison wallows in a lowland tallgrass prairie. The 6th Triennial American Bison Society Conference and Workshop, 28 October to 2 November 2019, Pojoaque, NM, USA, 25 pp.

Cochner, J., and D. Jenson. 1981. 1980 Mormon Island Crane Meadows fish inventory. Report submitted to The Nature Conservancy, Grand Island, Nebraska, USA, 36 pp.

Collins, S.L., and L.B. Calabrese. 2012. Effects of fire, grazing and topographic variation on vegetation structure in tallgrass prairie. Journal of Vegetation Science 23(3):563-575.

Currier, P.J. 1982. The floodplain vegetation of the Platte River: phytosociology, forest development, and seedling establishment. Dissertation, Iowa State University, Ames, Iowa, USA, 332 pp.

Currier, P.J., and C.A. Davis. 2000. The Platte as a prairie river: a response to Johnson and Boettcher. Great Plains Research 10:69-84.

Davis, C.A. 2005. Breeding Bird Communities in Riparian Forests along the Central Platte River. Great Plains Research 15:199-211.

Geluso, K., B.T. Krohn, M.J. Harner, and M.J. Assenmacher. 2013. Whooping cranes consume plains leopard frogs at migratory stopover sites in Nebraska. The Prairie Naturalist 45:91-93.

Hart, R. 2001. Where the Buffalo Roamed – Or Did They? Great Plains Research 11:83-102.

Hengst, G.E., and J.O. Dawson. 1994. Bark properties and fire resistance of selected tree species from the central hardwood region of North America. Canadian Journal of Forest Research 24(4):688-696.

Horncastle, V.J., E.C. Hellgren, P.M. Mayer, A.C. Ganguli, D.M. Engle, and D.M. Leslie Jr. 2005. Implications of Invasion by *Juniperus virginiana* on Small Mammals in the Southern Great Plains. Journal of Mammalogy 86(6):1144-1155.

Jones, S.M, R.E. Ballinger, and J.W. Nietfield. 1981. Herpetofauna of Mormon Island preserve. Hall, County, Nebraska. Prairie Naturalist 13:33-41.

Johnson, C., and S.E. Boettcher. 2000. Wooded or Prairie River? Great Plains Research 10:39-68.

Junk, W.J., P.B. Bayley, and R.E. Sparks. 1989. The flood pulse concept in river-floodplain systems. Canadian special publication of fisheries and aquatic sciences 106(1):110-127.

Kantrud, H.A., G.L. Krapu, and G.A. Swanson. 1989. Prairie basin wetlands of the Dakotas: a community profile. Biological Report (No. 85), U.S. Fish and Wildlife Service, Washington, DC, USA, 116 pp.

Kaul, R.B., D. Sutherland, and S. Rolfsmeier. 2012. The flora of Nebraska, 2nd edition. School of Natural Resources, University of Nebraska-Lincoln, Lincoln, NE, USA.

Keddy, P.A. 2010. Wetland Ecology: Principles and Conservation, 2nd edition. Cambridge University Press, Cambridge, UK, 497 pp.

Kolstad, O.A. 1981. A preliminary survey of the vascular flora of Mormon Island Crane Meadows, Hall County. Kearney State College Herbarium, Kearney State College, Kearney, Nebraska, USA, 58 pp.

Ladybird Johnson Wildflower Center (LBJWFC). 2022. Vegetation Surveys. University of Texas at Austin, Austin, Texas, USA. <https://www.wildflower.org/our-work/research>

Lingle, G.R., and M.A. Hay. 1982. A checklist of the birds of Mormon Island Crane Meadows. Nebraska Bird Review 50:27-36.

Long Term Ecological Research Network (LTER). 2018. About the LTER Network. National Science Foundation, LTER Network Office, Santa Barbara, California, USA. Retrieved at http://www.lternet.edu/network

Minnesota Department of Natural Resources (MN DNR). 2016. Minnesota's Wildlife Action Plan 2015-2025. Division of Ecological and Water Resources, MN DNR, St. Paul, Minnesota, USA, 240 pp.

Nagel, H.G. 1981. Vegetation and pollination ecology of Crane Meadows. Report submitted to The Nature Conservancy, Grand Island, Nebraska, USA, 80 pp.

National Ecological Observatory Network (NEON). 2020. Data & Samples. Battelle, Columbus, Ohio, USA. < https://www.neonscience.org/data-samples>

Noss R.F., E.T. LaRoe, J.M. Scott. 1995. Endangered ecosystems of the United States: a preliminary assessment of loss and degradation. Biological Report (No. 28), U.S. Department of the Interior, National Biological Service, Washington, DC, USA, 65 pp.

Powell, A.F.L.A. 2006. Effects of prescribed burns and bison (*Bos bison*) grazing on breeding bird abundances in tallgrass prairie. The Auk 123(1):183-197.

Powell, D.S. 2000. Forest Service framework for inventory and monitoring. White Paper for the Washington Office Ecosystem Management Corporate Team & Interregional Ecosystem Management Coordinating Group, U.S. Department of Agriculture, Forest Service, Washington, D.C., USA, 12 pp.

Ratcliffe, B.C. 1981. A preliminary inventory of the insects of Mormon Island refuge. Report submitted to The Nature Conservancy, Grand Island, Nebraska, USA, 22 pp.

Ratcliffe, B.C., P.C. Hammond. 2002. Insects and the native vegetation of Nebraska. Transactions of the Nebraska Academy of Sciences 28:29–47

Rolfsmeier, S.B., and G. Steinauer 2010. Terrestrial ecological systems and natural communities of Nebraska (Version IV). Nebraska Natural Heritage Program, Nebraska Game and Parks Commission, Lincoln, Nebraska, USA, 228 pp.

Rosas, C.A., D.M. Engle, J.H. Shaw, and M.W. Palmer. 2008. Seed Dispersal by *Bison bison* in a Tallgrass Prairie. Journal of Vegetation Science 19(6):769-778.

Schroeder, K., Assistant State Private Lands Coordinator, U.S. Fish and Wildlife Service. Contacted May 2015. (308)-382- 6468 x 15, <kirk_schroeder@mail.fws.gov>

Springer, J.T. 1981. Abundance and Diversity of Mammals on the Crane Meadows of Mormon Island, Hall County, Nebraska. Report submitted to The Nature Conservancy, Grand Island, Nebraska, USA, 39 pp.

Steenhof, K., and M.N. Kochert. 1988. Dietary responses of three raptor species to changing prey densities in a natural environment. The Journal of Animal Ecology 57:37-48.

Steuter, A.A., and L. Hildinger. 1999. Comparative Ecology of Bison and Cattle on Mixed-Grass Prairie. Great Plains Research 9:329-342.

Tall Timbers Research Station and Land Conservancy. 2015. Research at Tall Timbers. Tallahassee, Florida, USA. <http://talltimbers.org/research-at-tall-timbers/>

Tiner, R.W. 2016. Wetland indicators: a guide to wetland formation, identification, delineation, classification, and mapping. CRC Press, Boca Raton, Florida, USA.

Toevs, G.R., J.J. Taylor, C.S. Spurrier, W.C. MacKinnon, and M.R. Bobo. 2011. Bureau of Land Management Assessment, Inventory, and Monitoring Strategy: For integrated renewable resources management. Bureau of Land Management, National Operations Center, Denver, Colorado, USA, 44 pp.

U.S. Bureau of Reclamation, U.S. Department of Interior. 1938. Aerial Imagery from July 4th, 1938, and November 19th, 1938. Rainwater Basin Joint Venture Image Library, Grand Island, Nebraska, USA.

U.S. Fish and Wildlife Service (USFWS). 2013. 7-Year Inventory and Monitoring Plan. USFWS
National Wildlife Refuge System. Natural Resource Program Center, Fort Collins,
Colorado, USA, 50 pp.

U.S. National Park Service (USNPS). 2015. Program Brief. Inventory and Monitoring Program,
Natural Resources Stewardship and Science, U.S. Department of Interior, Fort Collins,
Colorado, USA, 2 pp. <http://npshistory.com/publications/interdisciplinary/im/im-brief-
2015.pdf>

Wiese, J.D., and A.J. Caven. 2018. Dataset of the physical conditions of Green Ash (*Fraxinus
pennsylvanica*) in riparian woodlands along the central Platte River. Data in Brief.
https://doi.org/10.1016/j.dib.2018.10.063

Williams, G. P. 1978. The case of the shrinking channels: The North Platte and Platte Rivers in
Nebraska. Department of the Interior, Geological Survey Circular 781.

Chapter 1: Long-Term Inventory and Monitoring Plot Layout Design, Sampling, and Installation

Project Goals and Methods

The long-term inventory and monitoring plot design created a system of transects on which to base vegetation, avian, small mammal, and butterfly species of concern monitoring surveys. Special consideration was given to capturing the various vegetative communities present on the landscape when designing the transect layout (Nagel 1981; Currier 1982). We utilized a stratified random sampling design by creating polygons around the various ecotopes represented in each management unit (Naveh 1994; Coulloudon et al. 1999; Herrick et al. 2009). We then placed a GPS location in the center of each polygon and utilized a field randomization technique to assign a random starting point to each transect within those polygons. Transect bearings were randomly generated using a random number generator. We also generated extra transect bearings for cases where the assigned transect bearing did not match the landscape to be sampled (linear ridges and woodlots). We created the polygons by overlaying soil maps, land use history maps, and aerial imagery, while considering the various vegetative communities of the Platte River, topography, and flood frequency. We also incorporated site visits to various management units to assess and confirm pasture diversity and variability. We utilized Google Earth aerial imagery from 1993 to 2013, the Web Soil Survey's soil map data, historic aerial imagery from 1938 and 1998, and documents describing the land use history for all management units from the Crane Trust's files (U.S. Bureau of Reclamation 1938; Nagel 1981; Morton 2013; Harner and Morton 2014; Google Earth 2015; NRCS 2015; Crane Trust undated; Figure 2). As additional properties are acquired this process can be repeated to add monitoring plots to those areas.

Figure 2. Photos of Mormon Island (with additional graphics) from 1938 and 1981 demonstrating the land use history of Mormon Island, Hall County, Nebraska, USA.

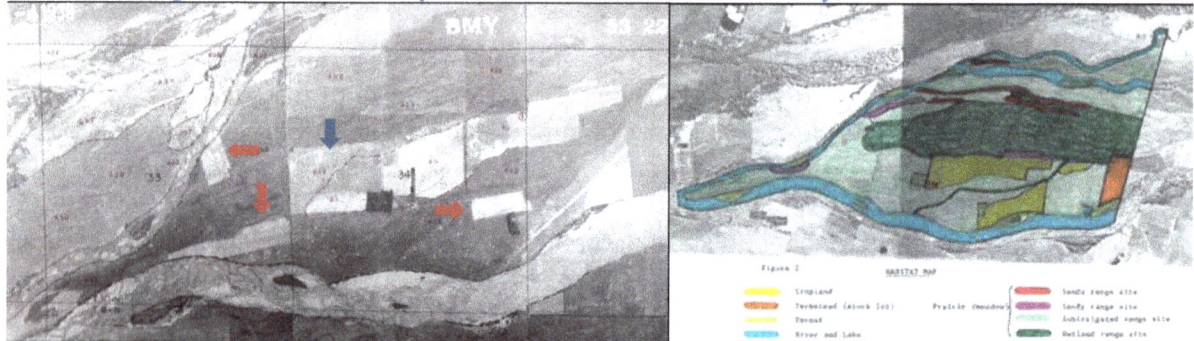

Notes: Sources- US Bureau of Reclamation 1938; Nagel 1981. The 1938 photos demonstrate some historic agricultural use not detected during 1980-1981 survey work. Areas of historic haying can be differentiated from cropland based on patterns and textures; red arrows indicate historic cropland (left) and a blue arrow indicates a hayed area (left). The photo on the right (Nagel 1981) has been color illustrated to describe the land uses as of 1981.

Sampling Design

Our major objective was to sample a robust and representative portion of the distinct ecological zones created by interactions of soil, land use history, and vegetative community types (ecotopes; Naveh 1994), with as much replication as possible given institutional constraints (labor, funding, etc.). Replication over space and time provides increased statistical and inferential power when answering ecological research questions (Elzinga et al. 2009; Figure 3). We selected soil types of interest based on their uniqueness within the landscape as well as the opportunity for relatively robust replication across space given the various vegetative communities and land use histories (NRCS 2015). We defined land use history as follows (See Caven et al. 2017): A "relict" prairie has never been tilled and to a great degree persists in its historic condition retaining remnant vegetation communities dominated by native species (relict components). A "restored" prairie has never been tilled but has been historically "over-utilized" or neglected to the degree that it lacked "a majority of relict components" and subsequent efforts have been made to return that pasture to its historic condition. Examples may include efforts to remedy historic chronic overgrazing, advanced woody encroachment, or historic exotic plant inter-seeding for livestock forage. "Reconstructed" prairie has been tilled, historically used for agriculture, and "subsequently seeded and replanted with native prairie species suspected to have inhabited that area in the past" (Caven et al. 2017).

Table 1. Monitoring plot breakdown

N = 74 Plots				
Habitat	**History**	**Flooding**	**Project Plots[2]**	**Dominant Soil Types[3]**

Prairie	36	Relict	47	Frequent	14	Avian	72	Platte-Bolent Complex	12
Wet Meadow[1]	18	Reconstructed	15	Occasional	30	Veg	69	Bolent-Calamus Complex	9
Savanna/Shrubland	8	Restored	7	Rare	17	SM	14	Wann Loam/Sandy Loam	8
Forest/Woodland	7	Open Water	5	Very Rare	8	BSOC	21	Inavale Loamy Sand	7
River/Pond	5			Permanent	5	Anuran	12	Gothenburg Loam	7
								Platte-Inavale Complex	5
								Calamus Loamy Fine Sand	5
								Barney Complex	4

Notes: 1) "Wet Meadows" defined by having a Wetland Indicator Score (WIS) of <3.1 from preliminary survey data. 2) "Avian" = avian point count surveys, "Veg" = vegetation quadrat and point-line intercept transects, "SM" = small mammal surveys, "BSOC" = butterfly species of concern surveys, and "Anuran" = anuran calling surveys, which are conducted on the indicated number of plots. 3) The "Dominant Soil Type" list is not exhaustive and includes only the most abundant soil types across plots (other = 12, open water = 5).

Figure 3. Aerial imagery of the long-term monitoring plot layout. AF1, AF2, DU1, DW1, HH1, HH2, HH3, and RBM3 are not pictured as they are not on the main complex. Imagery Google Earth (2020).

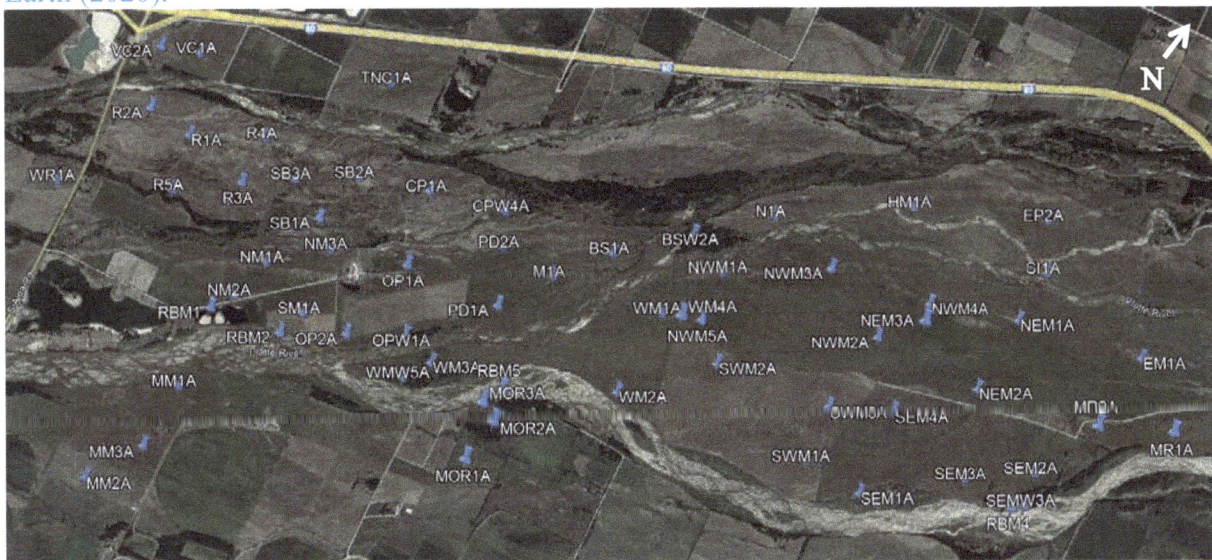

Our program straddles the line between a "monitoring" and "research" program as defined by Elzinga et al. (2009; Figure 3). Elzinga et al. (2009) contends that monitoring allows for the investigation of associations; for instance, Currier (1989) demonstrated that as ground water levels increased the cover of *Carex* spp. increased in wet meadow systems. Findings like this are highly informative, but don't include sufficient experimental control and replication to authoritatively demonstrate cause and effect relationships. Most ecological "research" projects that include sufficient experimental control and replication efforts provide additional inferential power but require a level of investment that is not sustainable for most small organizations. Contrastingly, "monitoring" efforts are by design long-term, but often do not allow for cause-and-effect inferences. Our program straddles the line between situation "D" and situation "E" in Figure 4 below. When management objectives, such as controlling woody encroachment or cool season exotic species, mandate spring-season controlled burns there are generally multiple comparable pastures within our long-term research and monitoring plot framework to allow both experimental replication and control. However, conducting field work takes resources. We are suggesting a minimum commitment of conducting vegetation and avian surveys on 50% of our 74 monitoring plots every year to track and investigate long-term trends. Additional resources given grants or other outside funding can allow us to conduct supplementary sampling and increase the robustness of experimental replication and control efforts. However, the baseline effort to complete vegetation and avian monitoring surveys on 50% of plots each year should allow us to effectively assess long-term trends regarding species composition in relation to management actions over time. The sampling framework also allows for replication and therefore improved experimental inference given appropriate survey effort.

Figure 4. Description of the continuum between "monitoring" and "research" from Elzinga et al. 2009. Our landscape-level biological research and monitoring program is broadly described by conditions "D" and "E" underlined below in red.

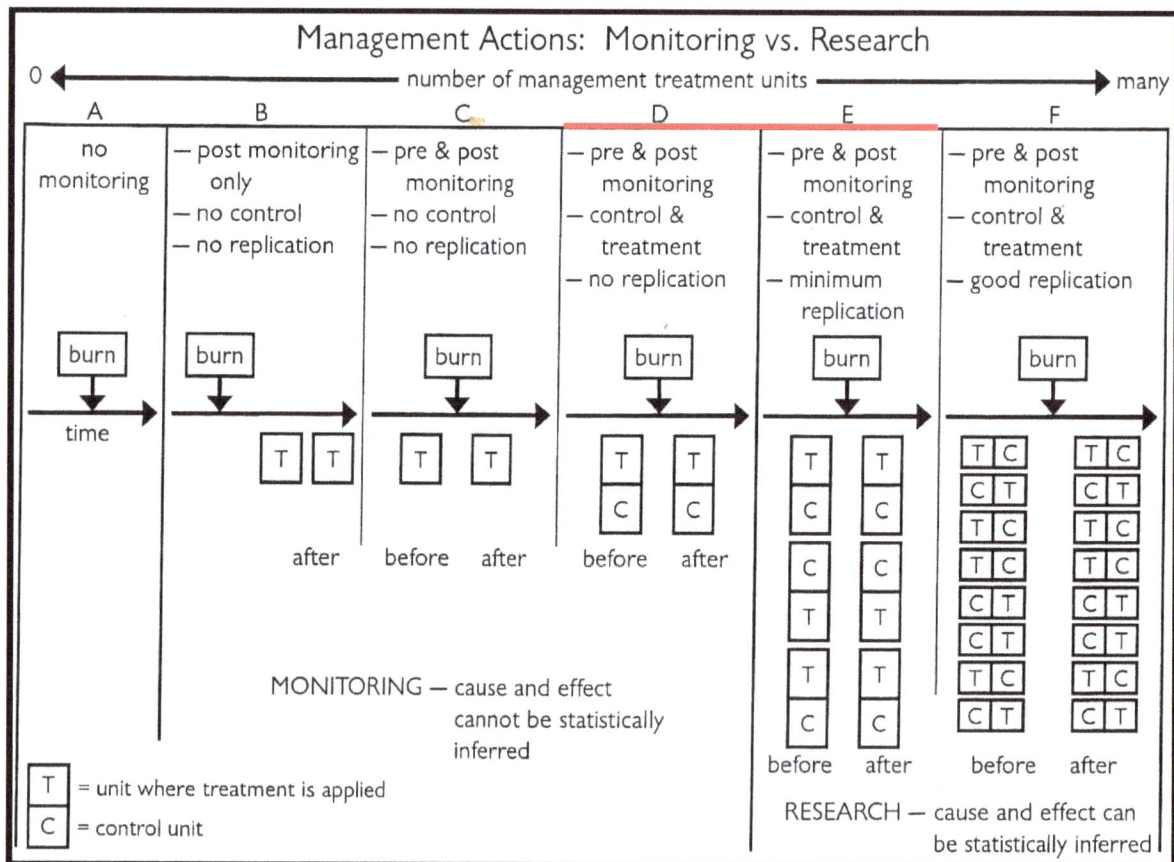

Instructions for Monitoring Plot Setup

The starting point was randomly determined by flipping a coin and throwing a ball from the initial GPS point placed directly in the center of each polygon delineating an ecotope of interest. The coin was flipped 3 times, and then a ball was thrown either lightly (3-5m) or somewhat harder (7-10m). The scheme looks like this:

> HHH= North Hard
> HHT= North Soft
> TTH= South Hard
> TTT= South Soft
> HTH= East Hard
> HTT= East Soft
> THH= West Hard
> THT= West Soft

A t-post was placed where the ball landed and a 100m tape measure was run out at the random bearing assigned to each monitoring plot. A capped rebar was placed at the end of the 100m transect. After placing the initial transect line, which will serve as the vegetation monitoring line, we moved back to the beginning of the transect (the t-post). We made a right angle to the bearing of the transect (90 degrees) and moved 10 meters to the right. We placed a capped rebar at this location. This will serve as the start of the small mammal, avian, and butterfly species of concern monitoring transect. We repeated the previous process and placed 1 more capped rebar at the end of the 100m transect, following the same bearing as utilized for the vegetation monitoring transect, creating a parallel transect 10 meters away. This serves to minimize the disturbance on the vegetation transect from repeated surveying. Each plot will contain 1 t-post, and 3 capped rebar stakes. When conducting point-line intercept and quadrat surveys on the vegetation transect always walk on the right-hand side of the transect and place the quadrats on the left-hand side of the transect. Small mammal traps will be placed directly on the transect 10m to the right. Please take a GPS point for all capped rebar placed in the ground, labeling the points as the transect ID, "SEM1" for example, then the letters "a" through "d". An example is drawn below.

Figure 5. Diagram of monitoring transect

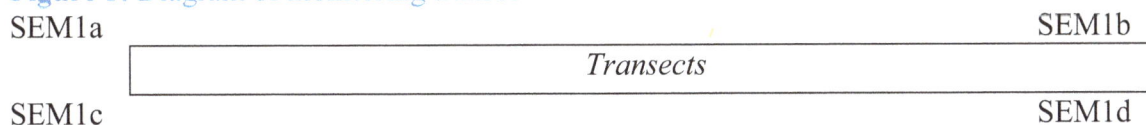

SEM1a   SEM1b

Transects

SEM1c   SEM1d

References

Caven, A.J., K.C. King, J.D. Wiese, and E.M.B. Buckley. 2017. A descriptive analysis of Regal Fritillary (*Speyeria idalia*) habitat utilizing biological monitoring data along the big bend of the Platte River, NE. Journal of Insect Conservation 21:183–205.

Coulloudon, B., K. Eshelman, J. Gianola, N. Habich, L. Hughes, C. Johnson, M. Pellant, P. Podborny, A. Rasmussen, B. Robles, P. Shaver, J. Spehar, and J. Willough. 1999. Sampling Vegetation Attributes: Interagency Technical Reference. Cooperative Extension Service: U.S. Forest Service- Department of Agriculture, Grazing Land

Technology Institute- Natural Resource Conservation Service, Bureau of Land Management- U.S. Department of the Interior.

Currier, P.J. 1989. Plant species composition and groundwater levels in a Platte River wet meadow. Proceedings of the Eleventh North American Prairie Conference 11:19-24.

Crane Trust. Undated. Mormon Island Wild Rose Historical Management Units Map.

Currier, P.J. 1982. The floodplain vegetation of the Platte River: phytosociology, forest development, and seedling establishment. Dissertation. Iowa State University, Ames, Iowa, USA, 332 pp.

Elzinga, C.L., D.W. Salzer, J.W. Willoughby, and J.P. Gibbs. 2009. Monitoring plant and animal populations: a handbook for field biologists. John Wiley & Sons, Hoboken, New Jersey, USA.

Grossman, D.H., D. Faber-Langendoen, A.S. Weakley, M. Anderson, P. Bourgeron, R. Crawford, K. Goodin, S. Landaal, K. Metzler, K.D. Patterson, M. Pyne, M. Reid, and L. Sneddon. 1998. International classification of ecological communities: terrestrial vegetation of the United States. Volume I. The National Vegetation Classification System: development, status, and applications. The Nature Conservancy, Arlington, Virginia, USA.

Nagel, H.G. 1981. Vegetation and pollination ecology of Crane Meadows. Report submitted to The Nature Conservancy, Grand Island, Nebraska, USA, 80 pp.

Naveh, Z. 1994. From Biodiversity to Ecodiversity: A Landscape- Ecology Approach to Conservation and Restoration. Restoration Ecology 2(3):180-189.

Google. 2015. Google Earth (Version 7) [Computer program]. Searching historic aerial imagery from 1993-2013.

Harner, M., and M. Morton. 2014. Copy of Crane Trust Total Acre Counts. Crane Trust Internal Document, Microsoft Excel File. Last Updated 7/24/2014.

Herrick, J.E., J.W. Van Zee, K.M. Havstad, L. M. Burkett, and W. G. Whitford. 2009. Monitoring Manual for Grassland, Shrubland and Savanna Ecosystems: Volumes I and II. USDA – ARS Jornada Experimental Range, Las Cruces, New Mexico, USA.

Morton, M. 2013. Property Information Document. Crane Trust Internal Document, Microsoft Excel File. Last Updated 3/1/2013.

Natural Resources Conservation Service (NRCS), United States Department of Agriculture. 2015. Web Soil Survey. Maps for Hall County, Nebraska, USA. <http://websoilsurvey.nrcs.usda.gov/>, Accessed 04/15/2015- 05/05/2015.

U.S. Bureau of Reclamation, US Department of Interior. 1938. Land Classification Maps: Imagery from July 4th 1938 and November 19th 1938 Aerial Surveys. Rainwater Basin Joint Venture Image Library, Grand Island, Nebraska, USA.

Appendix 1. Individual monitoring plot information and survey rotation

Tran. ID	Lat. (N)	Long. (W)	Deg.	Veg. & Avian	SM	BSOC	Soils	Flooding NRCS, (2015-17)	Relict/Restored	Habitat
AF1	40.761390	-98.509700	195	Even†			Wann Loam	Rarely, (no)	Reconstructed, 2009	Prairie
AF2	40.757530	-98.510700	135	Odd			Janude Loam, Calcareous	Rarely, (yes)	Reconstructed, 2009	Prairie
AF3	40.762831	-98.503104	257	Even		X	Wann Loam	Rarely, (yes)	Reconstructed, 2019	Wet Meadow-Prairie Transition
BS1	40.797464	-98.444065	116	Even†	X		Platte-Inavale Complex, 0-6% slope	Occasionally, (no)	Relict	Prairie
CP1	40.795730	-98.462120	318	Even	O	X	Bolent-Calamus/ Platte-Bolent Transition	Occasionally, (no)	Restored, tree removal, overgrazed, never tilled	Prairie-Savanna

Plot	Latitude	Longitude		Even/Odd			Soil	Flooding	History	Habitat
DU1	40.708077	-98.788742	151	Even†			Platte Soils (course sandy loam)	Somewhat Rarely, (no)	Restored, tree removal, never tilled	Prairie-Savanna (ridge)
DW1	40.704988	-98.786473	340	Odd†			Gothenburg Loam	Frequently, (yes)	Reconstructed, tree removal, disked and reseeding	Wet Meadow
EM1	40.807125	-98.395013	308	Even			Inavale Loamy Sand, 3-11% slope	Very Rarely, (no)	Relict	Prairie (ridge)
EP2	40.812974	-98.408179	208	Odd			Bolent-Calamus Complex	Somewhat Frequently, (yes)	Relict	Wet Meadow
HH1	40.774409	-98.515411	122	Even			Barney Complex	Frequently, (yes)	Relict	Wet Meadow
HH2	40.776737	-98.530193	50	Odd		X	Platte-Inavale	Occasionally, (no)	Relict	Prairie
HH3	40.774188	-98.522354	110	Even		X	Alda Loam-Inavale Loamy Sand	Rarely-Very Rarely, (no)	Reconstructed, 2019	Prairie (former Alfalfa)
HM1	40.809659	-98.420250	50	Even			Platte-Bolent Complex	Somewhat Frequently, (yes)	Relict	Wet Meadow
M1	40.794158	-98.448223	149	Even	X	X	Bolent-Calamus Complex	Occasionally, (yes)	Relict	Wet Meadow
MM1	40.774356	-98.476455	214	Odd†			Platte-Bolent Complex	Occasionally, (yes)	Relict	Wet Meadow
MM2	40.767621	-98.479616	273	Even			Alda Loam	Rarely, (possibly)	Relict	Prairie
MM3	40.771233	-98.475862	303	Odd			Calamus Loamy Fine Sand-Alda Loam Transition	Rarely, (no)	Relict	Prairie
MR1	40.802650	-98.38955	40	Odd		X	Wann Loam	Rarely, (no)	Reconstructed 2020 (from soybean-corn rotation)	Prairie
MR2	40.80072	-98.39617	70	Even			Wann Sandy Loam	Rarely, (no)	Reconstructed 2020 (from alfalfa)	Prairie
N1	40.804736	-98.431828	221	Odd†			Platte-Inavale Complex	Occasionally, (no)	Relict	Prairie
NEM1	40.805780	-98.406810	20	Odd			Bolent-Calamus Complex	Occasionally, (yes)	Relict	Prairie (mesic, ridge)
NEM2	40.800215	-98.407738	61	Even†	X		Barney Complex	Frequently, (yes)	Relict	Wet Meadow
NEM3	40.801950	-98.414830	96	Even			Barney Complex	Frequently, (yes)	Relict	Wet Meadow
NM1	40.786254	-98.473096	270	Odd		X	Inavale Loamy Sand, 3-11% slope	Very Rarely, (no)	Relict	Prairie
NM2	40.783206	-98.474458	258	Odd			Platte-Bolent Complex	Occasionally, (yes)	Relict	Wet Meadow
NM3	40.789021	-98.468244	249	Even			Calamus Loamy Fine Sand	Rarely, (yes)	Relict	Prairie
NWM1	40.799614	-98.434174	171	Odd			Platte-Bolent Complex	Occasionally, (yes)	Relict	Wet Meadow
NWM2	40.800324	-98.418266	309	Odd†			Barney Complex	Frequently, (yes)	Relict	Wet Meadow
NWM3	40.803220	-98.424935	50	Even			Platte-Bolent/Platte-Inavale Transition	Occasionally, (yes)	Relict	Wet Meadow-Prairie Transition
NWM4	40.804039	-98.415125	300	Odd			Platte-Inavale Complex	Occasionally, (yes)	Relict	Prairie (mesic, ridge)
NWM5	40.795960	-98.434000	322	Even†	X		Barney-Bolent Complex	Frequently, (yes)	Relict	Wet Meadow-Shrubland
OP1	40.790388	-98.461061	347	Odd	O	X	Calamus Loamy Fine Sand	Rarely, (yes)	Restored (ongoing), seeded with exotic grasses, not tilled	Prairie (mesic)
OP2	40.784258	-98.463378	32	Even†			Bolent-Calamus Complex	Occasionally, (yes)	Restored (ongoing), seeded with exotic grasses, not tilled	Prairie (mesic)
PD1	40.790665	-98.451714	69	Odd			Platte-Bolent Complex	Occasionally, (yes)	Relict	Wet Meadow
PD2	40.794461	-98.453664	121	Even		X	Inavale Loamy Sand, 3-11% slope	Very Rarely, (no)	Relict	Prairie
R1	40.791778	-98.484778	340	Even	X	X	Inavale Loamy Sand, 3-11% slope	Very Rarely, (no)	Relict	Prairie
R2	40.792309	-98.489263	165	Odd		X	Bolent-Calamus Complex	Occasionally, (no)	Relict	Prairie
R3	40.790510	-98.478410	31	Even		X	Platte-Bolent Complex	Occasionally, (no)	Relict	Prairie-Shrubland Transition
R4	40.793953	-98.478237	330	Odd		X	Inavale Loamy Sand, 3-11% slope	Very Rarely, (no)	Relict	Prairie
R5	40.787583	-98.484167	275	Even†	X	X	Bolent-Calamus Complex	Occasionally, (yes)	Reconstructed, 1998, 60-acre cornfield	Prairie
SB1	40.790630	-98.470260	253	Even	X	X	Platte-Bolent Complex	Occasionally, (no)	Relict, woody inv., some tree removal	Shrubland
SB2	40.794370	-98.468770	218	Odd	X	X	Inavale Loamy Sand, 3-11% slope	Very Rarely, (no)	Relict, woody inv.	Prairie
SB3	40.792330	-98.474140	16	Odd†		X	Platte-Bolent Complex	Occasionally, (yes)	Relict, woody inv.	Shrubland
SEM1	40.790063	-98.413953	227	Odd†	X		Bolent Fine Sandy Loam	Occasionally, (yes)	Relict, S of "Mormon Field 11" reconstructed, 1982	Shrubland

SEM2	40.796741	-98.399814	298	Even			Caruso Loam	Rarely, (no)	Reconstructed, 2017, historically "Mormon Corral"	Prairie
SEM3	40.794194	-98.405608	325	Odd	X	X	Caruso Loam	Rarely, (no)	Relict, Formerly "Mormon Field 8"	Prairie
SEM4	40.796244	-98.414142	44	Even†	X	X	Wann Sandy Loam	Rarely, (no)	Reconstructed, 1981, Formerly "Mormon Field 7"	Prairie
SI1	40.809643	-98.406227	78	Odd			Calamus Loamy Fine Sand	Rarely, (yes)	Relict	Prairie
SM1	40.784215	-98.467895	123	Odd			Platte-Bolent Complex	Occasionally, (yes)	Restored, 1993, altered hydrology to increase flooding	Wet Meadow-Prairie Transition
SWM1	40.790404	-98.419859	335	Even			Bolent Loam	Occasionally, (yes)	Relict, Formerly "Mormon Field 10", Never Tilled	Prairie
SWM2	40.793702	-98.430981	166	Even			Wann Sandy Loam	Rarely, (no)	Relict, Formerly "Mormon Field 6", consistently hayed (photos 1938, 1993, & 1998)	Prairie (mesic)
SWM3	40.794492	-98.419900	305	Odd			Wann Loam	Rarely, (no)	Reconstructed, 1981, Formerly "Mormon Field 7"	Prairie
TNC1	40.801309	-98.469961	74	Odd			Platte-Bolent Complex	Occasionally, (yes)	Relict (at site of plot), possible reconstruction NW corner of pasture	Prairie
VC1	40.797335	-98.487389	104	Even			Wann Loam	Rarely, (no)	Reconstructed, 2008	Prairie
VC2	40.796542	-98.491049	29	Odd			Inavale Loamy Fine Sand	Very Rarely, (no)	Reconstructed, 2008	Prairie
WM1	40.795174	-98.437645	311	Even			Bolent Loam	Occasionally, (yes)	Relict	Prairie (mesic)
WM2	40.788943	-98.438225	277	Odd		X	Barney-Bolent Complex	Frequently, (no)	Restored, cropped in 1938 aerial photos. Possibly natural recolonization of native plants.	Prairie
WM3	40.785076	-98.455059	257	Even†			Bolent-Calamus Complex	Occasionally, (yes)	Relict	Shrubland (mesic)
WM4	40.794784	-98.435756	354	Odd§			Barney-Bolent complex	Frequently, (Yes)	Restored 2019	Wet Meadow
WR1	40.784833	-98.494364	254	Odd	X	X	Platte-Inavale Complex, 0-6% slope	Occasionally, (yes)	Relict, tree removal, never tilled (N 1/2, S 1/2 was tilled)	Prairie
MOR1	40.77938	-98.4483	248	Even			Platte-Bolent	Occasionally, (yes)	Relict	Wet Meadow
MOR2	40.78254	-98.44743	56	Odd			Calamus Loamy Fine Sand	Rarely, (unk)	Relict	Woodland
MOR3	40.78337	-98.44906	88	Even			Bolent-Calamus	Occasionally, (unk)	Relict	Prairie
OPW1	40.785987	-98.458373	299	Even*			Gothenburg Loam	Frequently, (yes)	Relict	Woodland
BSW2	40.801144	-98.437979	78	Odd*†			Gothenburg Loam	Frequently, (yes)	Relict	Forest
SEMW3	40.794402	-98.399457	25	Even*†			Gothenburg Loam	Frequently, (yes)	Relict	Forest-Woodland
CPW4	40.796780	-98.454930	25	Even*†			Gothenburg Loam	Frequently, (no)	Relict	Woodland
WMW5	40.783258	-98.456950	50	Odd*†			Gothenburg Loam	Frequently, (yes)	Relict	Forest
MORW6	40.784470	-98.447480	71	Odd*§			Gothenburg Loam	Frequently, (yes)	Relict	Forest
RBM1	40.781742	-98.475853	NA	Even†			NA	NA	NA	Pond
RBM2	40.782322	-98.469050	NA	Odd†			NA	NA	NA	River
RBM3	40.762647	-98.509461	NA	Even†			NA	NA	NA	River
RBM4	40.793912	-98.400413	NA	Odd†			NA	NA	NA	River
RBM5	40.7847	98.44798	NA	Even†			NA	NA	NA	River

Notes: "*" Indicates that it is only necessary to read vegetation at woodlands following management actions. "†" Indicates this site would be good for avian monitoring during the winter and/or migration as a result of its accessibility and preliminary data. "§" Notes that only vegetation surveys are completed on a rotational basis (i.e., no avian point counts at survey location). "SM" signifies small mammal monitoring and "BSOC" signifies butterfly species of concern monitoring. "X" signifies survey done annually and "O" indicates survey is optional given staff time. "Odd" and "Even" refer to surveys being conducted in years ending in an odd or even number. Habitat classifications at monitoring sites are based on initial site visits and assessment via aerial imagery following Currier (1982) and Grossman et al. (1998) as well as plant wetland indicator statuses from initial surveys. Habitat classifications are updated following relevant management actions (tree clearing, etc.).

Chapter 2 Part A: Vegetation Monitoring Protocol

Project Goals

The vegetation monitoring protocol seeks to implement an easy to utilize standard method for monitoring land management impacts on vegetative communities over time. It also seeks to understand longer-term changes in the vegetation community that result from vegetative succession and global climate change. The only way to detect these long-term changes and the effects of management is a long-term vegetation monitoring strategy that considers soils, management units, land use history, and the vegetative succession processes (See Currier 1982; Naveh 1994). The desire to gather specific data that is sensitive to these changes must be balanced with pragmatic considerations, such as time and financial budgets. Each of the 69 vegetation transects are to be read 1 time every two years at a minimum. They should also be read each year after a controlled burn from the preceding spring, previous fall, or previous summer. Surveys are to be conducted from about early July to late September based on initial survey data, which suggests these months provide a relatively accurate depiction of vegetation biomass (as indicated by height class) as well as a good representation of the floral community in a relatively mature state, leading to higher identification accuracy (Caven and Wiese 2022). This survey season is relatively long and flexible by design as it allows for the consideration of environmental conditions. In wet years surveys can often be conducted into early October, but in dry years it may even be necessary to start in late June and complete surveys by early August to get an estimate of peak biomass and accurately identify plants (i.e., plants become desiccated earlier in dry years). Though early July (~late June) to late September (~early Oct.) is the recommended survey season, these transects can be read outside of this period to answer particular questions. For instance, estimating vegetative cover in Rough-legged Hawk wintering territories. However, this data should not be stored within the vegetation-monitoring database, but elsewhere pertaining to that specific study.

Project Methods

To monitor vegetative community changes, we use two primary vegetation monitoring techniques including the point-line intercept (also referred to as the "line-point intercept") method and the quadrat ocular cover estimation method. These techniques have been highly tested and represent the two major sampling methods for assessing changes in vegetation over time. The reason we utilize both methods is that they excel at collecting different types of data, though they overlap in the information they collect. The point-line intercept method excels at gathering data on vegetative cover and the dominant plant species. It is also easy to analyze and fairly quick to collect (Symstad et al. 2008; Herrick et al. 2009). However, this method does not as effectively capture species richness, which is of increasing importance as land managers refocus their priorities from managing for production to managing for diversity. The quadrat ocular cover estimation method consistently detects more species than does the point-line intercept method (Symstad et al. 2008). However, this method results in more variation in the mean percent cover estimate between observers; nevertheless, with proper training the cover estimation procedure can become quite standardized and accurate between observers (Symstad et al. 2008). For this reason, we employ both methods, which should give us the ability to detect rather slight changes in species composition and cover. Having two methods of cover can help us in two major ways. First, we will be able to confirm slight changes in cover and species composition by consulting both models. If they both demonstrate the same trend, we can be more certain in our conclusions. Secondly, using both methods will provide more specific cover data from more sparsely distributed plant species (quadrat) while providing optimal and easily quantifiable general cover data (point-line intercept). Caven and Wiese (2022) recently published a vascular plant inventory of Crane Trust properties using these methods and determined that 549 species have been detected locally since research efforts began in the early 1980s.

Point-Line Intercept

A 100m long transect will be permanently marked for each vegetation survey line. It is important to always walk on the left side of the transect from the starting t-post to minimize impacts on the site, try to walk at least a half meter away from the transect line at all times. A 100m tape measure will be laid out in the direction specified in the plot layout strategy for each transect. Before you start collecting data please have a data sheet, a clipboard, a species list, a pin flag, a tape measure, and of course a pencil.

At every 2 meters, starting at the 0.5-meter mark, record the vegetation by placing a pin flag straight down from the opposite side of the tape that you are on at the appropriate meter mark. Attempt not to press vegetation down with the tape measure, but instead get the tape as close to the ground as you can by working it into the vegetation. When recording dominant vegetation, think of an imaginary line running through the pin flag both up and down, record the dominant species that you intersect in each of the following height categories: short grass/forb, 0-0.5m; tallgrass/shrub, 0.5-2m (includes grasses over 2m); and subcanopy/canopy > 2m (woody species only) (SDN NPS 2019). If you find yourself intersecting 2 species in one or multiple levels, choose the dominant (apparently higher cover) plant from each level. If two plants are codominant in 2 strata, say *Panicum virgatum* and *Sorghastrum nutans* co-dominate the 0-0.5m and 0.5-2m strata, place one plant in each of the two categories to best represent the site. Consider how you have impacted the vegetation by placing the tape measure when appropriately putting vegetation into height categories. Finally, record the phenological state of the plant you are intersecting including the categories "vegetative" (V; during growing season before or after reproduction), "senesced" (S; dormant but still alive), flowering (FL; stamens and/or stigma produced), or fruiting (FR; fruit developing or mature). If the plant is both fruiting and flowering choose the predominant state (i.e., more flowers than fruit would equate to "FL"). Senesced plant material is nonliving, decadent, standing plant material that is still connected to a live plant, this is most commonly important when measuring grasses and other graminoids. Collecting this data can provide important information for controlled fire operations. Totally dead plants can be recorded as "snags" in the vegetation columns of the data sheet. Along with the 3 height categories of vegetation we also want to collect ground cover data. Where the pin flag hits the ground choose from the following categories when choosing a ground cover:

Plant Base - the base of a live plant
Bare ground - fine soils clean of obstruction
Rock/gravel - over 5mm in diameter
Moss/lichen - growing on the ground or rock
Litter/Duff - detached plant materials on ground in various levels of decay

Quadrats

We will be using 0.5m x 1.0m quadrats, marked in 10cm increments on the quadrat frame, to aid in the estimation of cover. Cover estimations will generally be made in increments of 5%. This is a modification of methods by Daubenmire (1959), Symstad et al. (2008), and Muldavin and Collins (2011). Overall cover data will be gathered from the point-line intercept method, the cover measurements gained from quadrat data will simply help us detect more subtle changes in species composition and relative abundance. Because of the overlaps in the cover of various species, and the rounded numbers, the cover estimates of this method will often exceed 100%, which is totally fine as long as observers are consistent. However, these numbers will also give us a good idea of compositional change in the vegetative community over time.

Daubenmire (1959) recommends estimating plant cover by creating a mental polygon around the outer edges of a plant within the quadrat; do not try to quantify the small interstitial spaces between the leaves of grass, forbs, and shrubs. This polygon should accurately and tightly fit around the outer edges of a plant or group of plants bunched together of one particular species, tally the total cover by species for each quadrat. Along with estimating plant cover, we would like to estimate the cover of exposed (not covered by a higher canopy) bare ground, rock or gravel, moss or lichen, and litter or duff on the ground's surface. For definitions of these ground covers refer to the list under the second paragraph of the point-line intercept heading. The polygon approach is also appropriate to these ground covers and refers simply to the ground not covered by plants in the lower canopy (0-0.5m).

To collect this data, place the 0.5m by 1m quadrat on the right side of the tape measure moving from the t-post start to the finish of the same 100m transect used for the point-line intercept data. You will walk up the left-hand side of the tape measure and record data every 10 meters, starting at meter 5 and continuing to meter 95. This will be a total of 10 quadrats. If staff resources become scarce or there are limited persons with botanical expertise working with the Crane Trust the quadrat portion of the vegetation survey could be completed on a longer rotation (every 3-5 years), but the point-line intercept data should be collected at least every two years at each site.

Photos

Physical monitoring efforts will be supplemented by keeping a photo log. Each time a site is visited for measurement a picture will be taken from 3 meters behind the t-post looking down the transect (pointing the direction of the transects' compass bearing). Please leave the tape measure extended while collecting photo data as it helps to highlight the transect in the photo. Also, compose the photo to capture as much of the landscape as possible, limiting the amount of pictured sky to about 10%. However, please include at least some sky as it helps to relocate the exact frame and conduct repeat photography.

Please download photos into the appropriate photo-monitoring folder for each plot on the Company (X) drive. Please save the photo as "TransectID_Date".

Data Management

Please scan all data sheets on the office copier and send them to jwiese@cranetrust.org. After scanning your raw data sheet, enter the data into the Microsoft Excel Database for each plot. There will be one database and data folder for each plot on the Company X drive in the

Vegetation Monitoring Folder. This will have subfolders for the different types of data under each plot folder (quadrat, point-line intercept, and photo). Databases for this project were created in Microsoft Excel and they will be constantly updated, after you enter your data simply press save. Databases are labeled by TransectID_Quadrat/Point_Database. These should be backed up on an external hard drive [or to a cloud server] at least 1 time per 2 weeks during data collection efforts.

Unknown Plants

Make every effort to identify plants in the field within reason. If you have spent over 45 minutes on one plant, it is about time to call it quits if you feel like you have not made much progress. If the plant is sufficiently abundant at the site, and you have determined it is not a species of concern, please collect and press this plant. Please list the dominant or co-dominant plants in the area, the area in which the plant was found (nearest transect/pasture), the date it was collected, its relative abundance, your initials, and a descriptive name for the plant. The descriptive name should be taxonomic in nature. For instance, if you have a Figwort-like plant, create a name to the appropriate taxonomic level you can certainly identify it to, proceeded by one adjective. For example, let's say the leaves were dissected and you knew it was in the Lamiales order for certain, and suspected it was a Figwort (Scrophulariaceae), but you could not be sure of that family. Give it a descriptive plant name of "Dissected Lamiales." For another example, let's say you actually are certain of the family, and are also pretty sure that it is in the type-genus of the family itself, *Scrophularia*; it also has hairy leaves. An appropriate label would be "Hairy *Scrophularia*." It is incredibly important that the label used on the datasheet to represent the unknown species corresponds directly to the initial label on the herbarium sample.

Herbarium Management, Catalog, and Labels Protocol
Purpose

This protocol is to serve anyone who may wish to add to, annotate, or create labels for cataloged specimens within the Crane Trust on-site herbarium. This three-part protocol is oriented for three main purposes: proper specimen management/handling, maintaining a cohesive and complete record of each mounted specimen, and ease of creating standardized labels for specimens.

New Specimen Management and Mounting

Plants brought in from the field should be unloaded from the field press and immediately placed in the large "LAB PRESS" between two sheets of newspaper and cardboard spacers and compressed for at least ONE FULL WEEK using a ratchet strap around the press. Collected specimens should each have the date, location, collector/identifier, habitat notes, and the species name (if known) written on bottom of the top sheet of newspaper. If the species name is unknown and the specimen was encountered on an official monitoring survey it should be given an identifier that corresponds directly to the data sheet from that survey. After the one week pressing time, specimens and their corresponding newspapers may be unloaded from the LAB PRESS and separated into three separate piles: READY TO MOUNT specimens (species scientific name known and ready to mount), UNK/CONFIRM ON PLOT (unknown species name, with importance directly related to monitoring and other specimens that need names confirmed), UNK/CONFIRM OFF PLOT (unknown species name of plants outside of monitoring plots and lower priority). Each pile will be noted on top with the pile name on a cardboard spacer. When an unknown (UNK) specimen is identified, it may be moved to the

READY TO MOUNT pile for mounting. County records should be identified by at least two competent staff and if the sample is sufficiently large it should be split and preserved in the Crane Trust herbarium and also sent to the University of Nebraska-Lincoln herbarium.

Specimens are to be mounted on acid-free herbarium paper using 1:1 water to Elmer's glue mixture. The preferred method of mounting specimens is by brushing out a thin layer of adhesive mixture over a glass pane, laying the pressed plant gently on the pane, delicately removing the specimen from the pane, and placing the plant across the herbarium paper. Other methods using a brush or glue bottle may be employed as well or coinciding with these methods. Specimens should be mounted with floral parts and key features displayed (if possible). Samples may need to be trimmed to fit within the bounds of the paper and a 3-inch tall by 4.5-inch-wide blank space should be left empty at the bottom right-hand corner to allow a space for the standardized label to be mounted. In this blank space, a note should be lightly written in pencil with the ID, location, and date to match samples with their labels. The corresponding newspapers should have the number of labels needed and then should be placed in a folder or box and saved until the sample records can be cataloged and a label is fitted to each glued specimen, then they may be recycled. Freshly glued samples are placed individually on a cardboard spacer and stacked with a sheet of parchment paper between each for 24 hours before removing the parchment paper and adding a label.

Herbarium Catalog

The yellow jump drive with the label "HERBA" will always contain the most updated records of herbarium specimens and property vegetation species list. The information from the newspaper of each mounted plant will be used to catalog samples placed into the herbarium on to the Excel document "Herbarium list(MASTER)". New specimens will be added to the bottom of the list, leaving 1 blank row below the bulk of the list and the newly added specimens. All information from newspaper should be included in the catalog under each corresponding data field (columns: Date, Family, Species, Common name, Location/paddock, Habitat/Notes, Collector, Identifier, and if possible, latitude and longitude). As each new specimen is added to the list, the Excel file "Vegetation(MASTER)SpeciesList_CraneTrust" should be referenced, adding newly detected species on the property to the list. New and old records on the property should be updated in each appropriate data field (Columns: Families, Code, Genus/species, Synonyms, Common Name, Wetland status (www.plants.usda.gov), Inventory (Last name of person who identified specimen and year first detected), Herbarium sample, Exotic/Native, County Record, Notes, Mormon, Shoemaker, and Plots with species). After a "set" (20-50 sample records) is transcribed from newspaper to the Herbarium catalog and all mounted specimens have been labeled, the blank row can be deleted, adding the new records to the bulk list.

Herbarium Labels

To save time and to keep labels standardized, the Excel to Word "Mail Merge" feature is used to create new labels. When a "set" of samples has been entered into the Herbarium List, all the new information from the columns and rows should be highlighted and copied as a unit. The second tab in the Herbarium List (Master) document titled "Mail merge" should be opened. The cell D2 (column D, row 2) should be selected by a single left-click. A right click in the same cell should open an options list and the "paste" button can be clicked, filling out the rest of the columns and rows. For each sample row, the "Island" column will need to be changed manually to reflect which island/property the collection was made on (Shoemaker Island, Mormon Island, Skinny

Island, Dipple Property, or Alda Farms). Once this is finished, save both Excel documents (Species and Herbarium MASTER lists) and close them. The Word document "Herbarium Labels Template Advanced" should then be opened. Upon opening, a notification screen will pop up indicating that a dataset has already been formatted and if that set should change, CLICK "YES" and open the HerbariumList(Master) Excel document in the pop-up window. The Word document will then bring all the rows in from the Herbarium "Mail Merge" tab into Word and input them into the formatted labels. Click the "Mailings" tab from the toolbars at the top of the Word window and click "Finish & Merge" button on the right side under the Mailing options tab. A dropdown window will appear and click on "Edit individual documents" button. At this point, the document is ready to print the labels, so under the File tab, click print. Close the advanced labels template document, but DO NOT SAVE!

The individual labels can then be cut from the printed pages and glued (using a glue stick) to their corresponding plant specimens, in the lower right-hand corner of the herbarium paper. Newly labeled specimens should be placed in a tote or box, labeled "Ready for the herbarium" to be put away in their respective folders. Folders are in alphabetical order by Family names. In each Family folder Genera are alphabetically separated into separate folders. Samples will be placed within their corresponding genus folder alphabetically by species, with newest samplings placed in front of older specimens. If either a Family or Genus folder is not found in the herbarium, it may be that a new one needs created, follow the identical filing scheme as the other folders and put away in the herbarium. Once all the mounted specimens are labeled, cataloged, and stored, their newspapers can be discarded in the recycle bin.

References

Caven, A.J., and J.D. Wiese. 2022. Reinventory of the Vascular Plants of Mormon Island Crane Meadows after Forty Years of Restoration, Invasion, and Climate Change. Heliyon 8:e09640.

Currier, P.J. 1982. The floodplain vegetation of the Platte River: phytosociology, forest development, and seedling establishment. Dissertation. Iowa State University, Ames, Iowa, USA, 332 pp.

Daubenmire, R.F. 1959. A canopy-coverage method of vegetational analysis. Northwest Science 33:43-64.

Herrick, J.E., J.W. Van Zee, K.M. Havstad, L.M. Burkett, and W.G. Whitford. 2009. Monitoring Manual for Grassland, Shrubland and Savanna Ecosystems: Volumes I and II. USDA – ARS Jornada Experimental Range, Las Cruces, New Mexico, USA.

Muldavin, E., and S. Collins. 2011. Core Research Site Web Quadrat Data for the Net Primary Production Study at the Sevilleta National Wildlife Refuge, New Mexico (1999-). Long Term Ecological Research Network, Albuquerque, New Mexico, USA.

Symstad, A.J., C.L. Wienk, and A.D. Thorstenson. 2008. Precision, Repeatability, and Efficiency of Two Canopy-Cover Estimate Methods in Northern Great Plains Vegetation. Rangeland Ecology and Management 61:419-429.

Sonoran Desert Network Inventory & Monitoring Program, National Park Service (SDN NPS). 2019. Uplands Monitoring in the Sonoran Desert Network and Chihuahuan Desert Network: Protocol Summary. National Park Service, Department of the Interior. <https://www.nps.gov/articles/uplands-monitoring-sonoran-desert.htm>

Appendix 2. Vegetation monitoring datasheets

Crane Trust Vegetation Monitoring – Quadrat Method   Date: _____   Transect ID: _____   Recorder: _____   Observer: _____   Pg. ___ of ___

Species/Ground	Quad. 5m	Quad. 15m	Quad. 25m	Quad. 35m	Quad. 45m	Quad. 55m	Quad. 65m	Quad. 75m	Quad. 85m	Quad. 95m

Ground Cover Classes: Plant Base (PB), Bare Ground (BG), Rock/Gravel (RG), Moss/Lichen (ML), Litter/Duff (LD)

Unknown Plants

Unknown Plant Descriptive Name: _____   Collection ID (MMDDYYYY_TransID_Quad_Number): _____

Plant Collected: Y or N   Pictures Taken: Y or N

Comments (Protocol): _____

Crane Trust Vegetation Monitoring - Point-Line Intercept Method

Date: _____ Transect ID: _____ Recorder: _____ Observer: _____

M	Veg. 0-5m	Veg 0.5-2.0m	Veg. 2m +	L/S	Ground	M	Veg. 0-5m	Veg. 0.5-2.0m	Veg. 2m+	L/S	Ground
1.5						51.5					
3.5						53.5					
5.5						55.5					
7.5						57.5					
9.5						59.5					
11.5						61.5					
13.5						63.5					
15.5						65.5					
17.5						67.5					
19.5						69.5					
21.5						71.5					
23.5						73.5					
25.5						75.5					
27.5						77.5					
29.5						79.5					
31.5						81.5					
33.5						83.5					
35.5						85.5					
37.5						87.5					
39.5						89.5					
41.5						91.5					
43.5						93.5					
45.5						95.5					
47.5						97.5					
49.5						99.5					

Ground Cover Classes: Plant Base (PB), Bare Ground (BG), Rock/Gravel (RG), Moss/Lichen (ML), Litter/Duff (LD); **L/S:** Live vs. Senesced Vegetation

Unknown Plants

Unknown Plant Descriptive Name: _____ Collection ID (MMDDYYYY_Trans ID_Point_Number): _____

Plant Collected: Y or N _____ Pictures Taken: Y or N _____

Comments (Protocol): _____

Comments (Land Management/Natural Disturbance): _____

30

Chapter 2 Part B: Visual Estimate of Grazing Impact for the Crane Trust Biological Monitoring Plan

Project Goals

Originally, we planned to summarize the unit and monitoring site management actions based on the Crane Trust's grazing and fire management plans and records alone. However, it became apparent following the first season of inventory, that given fencing issues and the multipronged (research, education, outreach, restoration, management, etc.) nature of the Crane Trust, that Cattle and Bison would regularly be grazing in unplanned locations. Therefore, it became imperative to develop a scale to visually estimate and quantify grazing impacts to corroborate the land management plan (fire, grazing, etc.). This scale will be used to substantiate the grazing management plan and assess the impact of stocking rates on a per-pasture and per-monitoring plot basis annually, as well as detect unplanned grazing activity. We will visually assess grazing impacts with every vegetation survey completed during biological monitoring efforts as well as conduct a post-growing season grazing assessment. This assessment should be done by examining the monitoring site previous to the start of data collection. Though the vegetative data itself, including vegetative height is a good indicator of grazing, it is important to corroborate via visual estimation to make sure the monitoring plan is highly standardized.

Project Methods

This scale is derived from work by Bruhjell and Moore (2003) and Kothmann and Hinnant (1993). It simply categorizes the level of grazing via a quick visual estimate. Kothmann and Hinnant (1993) use a 0–6-point scale, where "… zero represents no use and 6 represents total or extreme use. A rating of 3 represents full use. Full use is normally the maximum use desired on rangeland." Bruhjell and Moore (2003) recommends a scale of 1-5 for the same purpose, with a 1 representing "none-slight" grazing and a 5 representing "severe" grazing. Both scales indicate that a moderate level of grazing which includes the near uniform grazing of preferred plant species, with minimal impacts on subprime forage species and areas, to be an ideal level of grazing. Kothmann and Hinnant (1993) consider this level of grazing to be "full use" and Bruhjell and Moore (2003) describe this as "moderate" use. As a conservation organization we generally consider our ideal level of grazing to be right below full use and desire a rotation that creates patchy habitats ideal for many species of avifauna, small mammals, and insects. We also try to manage in a way that retains "thatchy" grasslands and some shrubland for those species adapted to such environments. We chose a scale of 0-5, which differentiates "no" and "slight" grazing. The scale more closely resembles Bruhjell and Moore's (2003) model, but both conceptual models are utilized. We enumerate our adapted scaler assessment below. Quotes around scale descriptions indicate that language closely reflects Bruhjell and Moore's (2003) and/or Kothmann and Hinnant (1993).

Table 2. Visual assessment of grazing level

- 0- None – "No detectible grazing"
- 1- Slight – "Ephemeral use, slight grazing of only most preferred species"
- 2- Light - "Only preferred areas and key forage species grazed"
- 3- Moderate - "Key areas grazed uniformly, especially key species"
- 4- Heavy - "Key species closely grazed and low forage value plants moderately grazed"
- 5- Severe - "Pasture appears mowed including low-value species"

Notes: Both scenarios 4 and 5 can promote weedy invaders, especially "increasers", such as Buffalo Bur (*Solanum rostratum*) which signifies disturbance and at least some bare ground.

References

Bruhjell, D., and T. Moore. 2003. Monitoring Riparian Areas. *Riparian Factsheet Series* (number 7 of 7). British Columbia Ministry of Agriculture, Food, and Fisheries, Kamloops, BC, Canada, 6 pp.

Kothmann, M.M, and R.T. Hinnant. 1993. The Grazing Manager: An Operational Level Grazing Management Decision Aid. Pages 140–147 *in* J.R. Cox and J.F. Cadenhead, editors, Symposia Proceedings of Project Range Care: Managing Livestock Stocking Rates on Rangeland. Department of Rangeland Ecology and Management, Texas Agricultural Extension Service, Texas A&M University, College Station, Texas, USA, 153 pp.

Chapter 3: Avian Monitoring Survey Protocol

Project Goals

This project is part of the broader Crane Trust biological monitoring plan. It seeks to gather data on the variety and relative abundance of bird species across various ecotopes and management regimes, throughout the year, and across several years, and decades. This project will create accurate and up to date bird species lists for each long-term monitoring transect, and the Crane Trust property as a whole, for each of the 4 seasons. The data will help us better understand the current spatial distribution and seasonal variation of the bird species utilizing the Crane Trust, and provide a great opportunity to pick up sightings of rare birds, either rarely sited or never previously documented at the Crane Trust. However, most importantly, this project will provide us with an opportunity for long-term trend analyses of avian populations on our properties and the ability to assess the impact of various management techniques on avian communities. Research is clearly showing a shift in bird migration and nesting patterns as a result of climate change, habitat fragmentation, and habitat change (Travis 2003, Opdam and Wascher 2004). It is a very important time in history to collect long-term data. Because of the incredible mobility of avian species, as compared to many other taxa, they often serve as one of the first indicators of climatic change and variation. Our biological monitoring plan allows us to evaluate the impacts of various land management practices, such as burning or haying frequency, the intensity of grazing, or the difference between Cattle and Bison grazing on bird communities (Fuhlendorf et al. 2009). Although this property has been extensively surveyed at different times in the past, this research protocol should allow us to update past survey results (Lingle and Hay 1982, Davis 2005). We will not be trying to estimate the real population based off the number of birds detected; we will simply count the number of birds detected and treat that as an index for discerning the relative abundance of particular species. We may divide these numbers into categories during the analysis stage.

Project Methods

From the summer of 2015 to the spring of 2018, we conducted 1/8 mile-long (200m) moving surveys along set transects to be completed in 15 minutes, and stationary focal point surveys to also be completed in 15 minutes (Sorace et al. 2000). This equated to 30 total minutes of effort per site via two survey methods. At riparian bird monitoring areas, we simply conducted 30-minute point counts. Sorace et al. (2000) demonstrated that the species richness, abundance, and diversity detected on point counts increased as survey time lengthened from 5 to 20 minutes. Gregory et al. (2004) described that point-counts are better at detecting avian communities in denser habitats such as shrublands, woodlands, and forests, and that transect counts are better at detecting species in open homogenous environments such as prairies or deserts. We decided that conducting both types of surveys in all terrestrial environments would be the best way to compile a relatively comprehensive record of the species using Mormon and Shoemaker Islands, and the surrounding landscape. Walking surveys tend to flush birds out of the grass that might not be easily detected without disturbance and point count surveys bring several birds closer to you as they realize you are not a threat (Bibby et al. 1992, Gregory et al. 2004). We employed both methods to maximize detection and maintain standardization. A two-person team, an expert "observer" who concentrates completely on the act of birding during surveys, and a "recorder" who documents data and navigates with the GPS device will conduct each survey effort. The recorder can help detect and identify birds as appropriate while completing their primary duties of navigation and data recording. However, counting should generally be the responsibility of the observer. It can be very helpful to tie a string around a light clipboard for the recorder so they can let the clipboard drop when they pick up their binoculars.

Each bird will be recorded as being detected within 50m or outside 50m of the transect or focal point (Gregory et al. 2004). We will attempt to accurately count large flocks, but in the case of large moving flocks (x>35) we can estimate to the nearest 5 birds. Age class and sex for individual birds will not be regularly collected during this study, but demographic notes of interest can be recorded in the database and notes section of the datasheet. Count every bird you see or hear while doing the point count, then figuratively "forget" that you ever saw them before embarking on your transect walk where you will again list every bird you see or hear. These can be listed on the same datasheet as there is a data entry space to record either "point or transect" detection. This data also lets us compare the effectiveness of each technique in various environments. Riparian Bird Monitoring sites are simply point counts conducted on wetlands with the help of a spotting scope, as visual distances may be significantly longer. No transects are included for these sites. If a transect is completed under 15 minutes simply continue to count birds from the end of the transect line until 15 minutes is reached. If 15 minutes is reached before the end of the transect, continue the survey until the end of the transect at an increased speed and record the time accurately in the database (for instance 17 minutes). If time is taken to identify a bird or chase a bird off transect for identification, the clock can be stopped and restarted upon the return to the transect. However, every effort should be made to keep surveys to the allotted 15-minute time periods to standardize effort across sites for habitat management modeling purposes. However, during the inventory phase individual surveys (transect counts, etc.) could be extended from 15 up to 30 minutes during migration periods to make sure all species were correctly identified and counted.

It is permissible to move off the line or focal point occasionally for the identification of a bird that is clearly detectable, but not identifiable from the focal point or transect line. For example, migrating *Ammodramus* spp. sparrows staying low on a relatively cold day. An observing team should move the minimal distance off the line necessary for the identification of the unknown species, once a positive identification has been made you will return to the transect line or focal point and continue the survey (you should stop the timer when you depart the focal point or transect for an extended period of time). This is especially important with potentially rare birds, for example, if a possible non-singing Hoary Redpoll was detected at about 55m off the line, but could not be identified with 100% certainty at that distance, the observer could and should wander off the transect line, or out of the focal point area for a closer look. If eventually the Hoary Redpoll is positively identified at a distance of 10m from the bird and 45m off the transect line, *the original detection distance should be recorded* (record as detected at over 50m on the data sheet). It should also be noted in the data sheet under "comments" that the observer walked 45m off the transect line for a positive identification. However, ensure that trampling of the vegetation line, typically 10 m to the left of the bird and small mammal line, is minimized. Leaving the focal point or transect line is only to be done within reason, some birds will simply be out of your detectable range, for example a small sparrow at 150m; you would be leaving the area of data collection. As a guideline try not to walk more than 50m off the transect line, and try to do it as infrequently as possible. The "comments" section of the database can be used in cases as detailed above, for describing detection methods, but it can and should also be used to record interesting avian interactions and foraging habits. Seeing a Northern Harrier grab a rodent or two Red-tailed Hawks collide in air, would be events of note for the "comments" section of the data sheet.

Every effort should be made to conduct surveys on days with good weather, which promotes the detection of bird species. As each site will only be visited twice per breeding season, it will be important to get as complete of a species list as possible during the site visits, which will represent the total species composition for that site and season in the data for that year. Therefore, weather that increases species detection is desired. Consulting other bird monitoring protocols, it is important that local ecological conditions be considered along with the goals of the project when deciding the appropriate weather conditions for a survey. An agreed upon set of factors diminishes the number of species and individual birds detected- rain, fog, cold (especially in conjunction with overcast skies), extreme heat, and wind.

For example, during a fall or winter survey, lower than average temperatures, in conjunction with high wind speed and overcast skies are very likely to reduce the detection of birds. It is important to consider all of these factors together before deciding to conduct a survey or not. For instance, even if it is cold, say 25° F, on a sunny day with low wind, wintering birds may still be perching and singing. However, in the same temperature with completely overcast skies and no possibility of thermal warming via feather surface area, the birds may be huddled under cover. Their energy use cost-benefit ratio will change, and it benefits them to seek shelter, decreasing detection. Basically, look outside and assess the general conditions before you go out for a survey, if you hear avian consistent activity, you are probably fine to do the survey. Here is a list of environmental factors to consider that reduce bird detection.

Table 3. Weather limits for bird surveys
> Weather below 28° F
> Wind above 15 mph
> Fully Overcast Skies
> Moderate to Heavy Rain

Weather above 90° F
Medium to Heavy Fog

A different set of conditions will be important in different seasons, with wind being the major factor in the spring, cold in the winter and fall, and heat in the summer.

Use of vocalizations

The use of recorded vocalizations is permissible in cases where it is needed to make a positive identification, either by using the recorded vocalization for comparison with what is being heard or to draw the individual of interest closer for visual confirmation. Using recorded vocalizations for callback purposes is not permitted during breeding season surveys because in some cases it can cause target breeding species to unnaturally leave their nests and expose themselves and their offspring to predation. Additionally, the use of recordings will influence detection probabilities for individual species. When using your electronic device to play bird calls to confirm them in the breeding season, play those calls quietly. It is also important to avoid using vocalizations to call forth birds when potential predator species (e.g., Cooper's Hawks) are known to be in the area, this is true regardless of season.

Data Management

All bird data will be recorded in the field using the American Union of Ornithologists Alpha Codes (Pyle and Desante 2014). These are 4-letter abbreviations of common names that are used in almost all professional bird surveys; it is quicker than writing the entire name or searching a checklist and is the absolute norm when consulting a variety of protocols (See referenced materials for various other protocols ours is modeled after). It also allows you to record any potential birds, especially rare ones that could show up unexpectedly, without needing to search a list or write the full common name. They follow a very simple logic, for instance, for a bird with a one-word name, such as a "Redhead", the code is the first 4 letters "REDH", for a bird with a 2-word name the code is simply the first 2 letters of each word, for instance Pine Siskin is recorded as "PISI," for a 3-word bird name it is usually the first letter of the first 2 words, and the first 2 letters of the final word, for instance "Red-winged Blackbird" is "RWBL", and as you can predict, birds with a 4-word name have codes made up of the first letter of every word, the Great Black-backed Gull has a code of GBBG. There is the rare occasion where 2 names overlap. For instance, several sparrows whose names start with a "Sa…", in the first word of the name. However, this is pretty rare and also resolved logically. In this case the first 3 letters of the first word are used and the first letter of the last word. Sage Sparrow, for instance, is "SAGS" and Savannah Sparrow is "SAVS." Most of the time I give it my best guess in the field and then double check with the Alpha Codes list before I enter the data into a computer. Alpha codes will be used in our database; however, you can write it differently on the field data sheet if you feel strongly about writing out part or all of the bird's entire name.

The data entry for this protocol is simple. Tally the total number of birds of a particular species seen outside or within 50 m of the survey point or transect, and insert that as a whole number into the appropriate data category on the Microsoft Access Database, as well as the correct Alpha Code for each species detected, whether the bird was observed during a point count or transect survey, any comments on behavior or data collection, the names of the observer and data recorder, the transect identification number, the date, the start and end times of the survey, the temperature, and the wind speed and direction. Observer's and recorder's names should be written as first initial, period, last name: "A. Caven" for example. Additionally, transects should be labeled "trans", point counts as "point", and incidental observations as "incidental" in the database. Weather data can be obtained from the internet after the survey but is better gained by using a Kestrel Weather Meter in the field if one is available. The database as well as field data forms (in Excel and PDF format), this protocol, and the bird species "Alpha Code" list, are located on the Company "X" Drive via the following pathway Science Program/Avian Monitoring. The Database is labeled "Avian Monitoring Survey Database" because of the way the database is set up you will see 2 possible icons to click on. Always click on the top icon, which also is a much bigger file. Please remember to save your data every time. Just go to the toolbar and press save. Datasheets, after entry, are to be scanned and emailed to bostrom@cranetrust.org. The datasheets are then stored in a filing cabinet in a particular folder within the Director of Conservation Research's office.

Project Update 2018

Our goal was to re-inventory the birds of Mormon Island about 35 years after the original inventory (Lingle and Hay 1982) as well as set a baseline for future monitoring efforts. Currently, we have moved past the re-inventory phase (*in-draft*) and into a long-term monitoring phase. We now simply conducted 15-minute point counts at all sites (including riparian sites), as this method is effective in examining the impact of various habitat parameters (i.e., percent vegetative cover, etc.) on avian communities (Gregory et al. 2004). During these surveys all detected species and the total number of each will be recorded (detected by sight and/or vocalization). Beginning in 2018, surveys have been conducted 2 times per breeding season (~21 May to 15 July) at each site every other year as well as at any site following recent management action (controlled burn, tree clearing, etc.). Additionally, a subset of sites representing a variety of habitats should be surveyed during each season throughout the year (winter: December - February, spring: March - May, summer: June - August, and fall: September - November). Specific sites for year-round monitoring could vary from year to year or include a permanent selection depending on research objectives. Currently, we have a set of priority sites to survey for during the spring and fall migrations as well as during the winter (Chapter 1, Appendix 1). However, we also sample outside of this list to answer particular questions (i.e., winter avian use of areas experiencing a fall controlled burn). We utilized data from initial surveys from June 2015 to June 2017 and data from Sharpe et al. (2001) to develop targeted survey ranges with which to document habitat use by various taxa and guilds during migration by averaging detection timings across two spring (2016 and 2017) and two fall (2015 and 2016) migrations via our data and including data from high, early, and late counts regionally by Sharpe et al. (2001; Table 4).

Table 4. Suggested survey timings to capture use by particular guilds and taxa. Statistical representation of dates uses Julian Day (day of year 1-365).

Group	Habitat	Mean First	SD Early	Mean Peak	SD Late	Mean Last	Proposed Survey Range	
			<-----	Spring	---->			
Shorebirds	Wetlands 1	82	103	127	151	158	4/15	5/21

36

Warblers	Woodlands	118	120	135	151	152	5/1	5/21
Sparrows	Grasslands	87	104	112	120	138	4/15	4/31
Ducks	Wetlands 2	38	60	80	100	134	3/1	4/10
			<-----	**Fall**	---->			
Shorebirds	Wetlands 1	190	213	229	245	294	8/5	9/1
Warblers	Woodlands	230	259	273	285	317	9/20	10/10
Sparrows	Grasslands	258	275	292	309	318	10/10	11/1
Ducks	Wetlands 2	272	288	316	343	357	10/25	12/1

References

Bibby, C.J., N.D. Burgess, and D.A. Hill. 1992. Bird Census Techniques. Academic Press, Inc. San Diego, CA.

Conway, C.J. 2011. Standardized North American Marsh Bird Monitoring Protocol. Waterbirds 34(3):319-346.

Davis, C. 2005. Breeding and migrant bird use of a riparian woodland along the Platte River in Central Nebraska. North American Bird Bander 30(3):109-114.

Fuhlendorf, S.D., D.M. Engle, J. Kerby, and R. Hamilton. 2009. Pyric herbivory: rewilding landscapes through the recoupling of fire and grazing. Conservation Biology 23(3):588-598.

Lingle, G.R., and M.A. Hay. 1982. A checklist of the birds of Mormon Island Crane Meadows. Nebraska Bird Review 50:27-36.

Opdam, P., and D. Wascher. 2004. Climate change meets habitat fragmentation: linking landscape and biogeographical scale levels in research and conservation. Biological Conservation 117:285–297.

Pyle, P., and D.F. DeSante. 2014. Four-letter (English Name) and Six-letter (Scientific Name) Alpha Codes for 2098 Bird Species (and 98 Non-Species Taxa) in accordance with the 55th AOU Supplement, sorted by English name. The Institute for Bird Populations. <www.birdpop.org>

Gregory, R.D., D.W. Gibbons, and P.F. Donald. 2004. Bird census and survey techniques. Pages 17-56 in W.J. Sutherland, I. Newton, and R.E. Green, editors, Bird Ecology and Conservation: A Handbook of Techniques. Oxford University Press, Oxford, United Kingdom.

Sharpe, R.S., W.R. Silcock, and J.G. Jorgensen. 2001. Birds of Nebraska: Their Distribution & Temporal Occurrence. University of Nebraska Press, Lincoln, Nebraska, USA, 520 pp.

Sorace, A., M. Gustin, E. Calvario, L. Ianniello, S. Sarrocco, and C. Carere. 2000. Assessing bird communities by point counts: repeated sessions and their duration. Acta Ornithologica 35(2):197-202.

Travis, J.M.J. 2003. Climate change and habitat destruction: a deadly anthropogenic cocktail. Proceedings of the Royal Society of London B: Biological Sciences 270:467–473.

Appendix 3. Avian point count and transect monitoring datasheet

Crane Trust- Avian Survey Form

Observer: _____   Recorder: _____   Date: _____   Temp: _____   Wind Speed/Direction: _____   pg. _____ of _____

Transect Number: _____   Start Time/ End Time: _____

Point/Trans.	Species	Inside 50 m	outside 50 m	Comments
1				
2				
3				
4				
5				
6				
7				
8				
9				
10				
11				
12				
13				
14				
15				
16				
17				
18				
19				
20				
21				
22				
23				
24				
25				
26				
27				
28				
29				
30				

38

Chapter 4: Small Mammal Monitoring

Project Goals

It is important to consider small mammals as a variable in a long-term monitoring plan for several reasons. They are very sensitive to landscape level ecological changes such as fire frequency, grazing activity, and woody encroachment (Collins 2000; Matlack et al. 2001; Horncastle et al. 2005; Johnston and Anthony 2006). Bison (*Bison bison*) grazing has been shown to positively impact deer mice (*Peromyscus maniculatus*) populations in the Great Plains in comparison to Cattle (*Bos tarus*) grazed sites and ungrazed sites (Matlack et al. 2006). It is notable, however, that in Matlack's 2006 study, which matched biomass between Cattle and Bison on grazing units, that both grazers had a positive impact on deer mice populations. In another study of wooded meadows in Oregon, Johnston and Anthony (2006) found that heavy Cattle grazing reduced small mammal populations and diversity when compared to light grazing. Successional changes can also impact small mammal populations. Horncastle et al. (2005) found that Eastern Red Cedar (*Juniperus virginiana*) encroachment upon grasslands in Oklahoma severely limited small mammal species diversity. Small mammals often serve as an indicator of landscape level changes, and they are an important prey item for raptors and other predatory wildlife. The National Science Foundation's LTER (Long Term Ecological Research) program has had several research sites across the country monitoring small mammal populations for decades. Monitoring small mammal populations has now become a standard practice for federal land management and stewardship agencies (USGS 2012; Newsome 2015). We made an effort here to follow suit.

Our goals are simple: first, we wanted to do a systematic inventory of small mammals across the various habitats and landscapes represented in our biological monitoring plan to get a better picture of how the various species present on the Crane Trust's core properties are distributed, and in what abundance (unpublished, completed 2015). Because of the great fluctuation in small mammal populations over time, it will be important to collect this data in perpetuity to understand the natural fluctuations in our small mammal populations and the effect of disturbance regimes such as flooding and management actions such as fire on our biological resources. Thus, our second goal will be to better understand the impact of our management actions on small mammal populations, and therefore make inferences about the potential prey base and habitat we are providing for predators such as mesocarnivores, birds of prey, and even Whooping Cranes and Sandhill Cranes, which are known to eat small mammals (Walkinshaw 1949; Allen 1952). We are now working to trap annually at 12-14 sites (2016 –), representing areas both grazed by bison and cattle in order to make long term comparisons. Sites were also selected based on initial trap success in 2015 in terms of sample species richness and total capture rates (Figure 6).

Project Methods

We use Sherman Box Traps baited with a birdseed mix to trap small mammals along our long-term monitoring transects. We elected to use a birdseed mix as bait because it has been found to be the most effective plant-based and easy to procure bait for trapping in the Great Plains (Oswald and Flake 1994). We use a custom mix of cracked corn (*Zea mays*, processed to improve its digestibility), black oil sunflower seed (*Helianthus annuus* - cultivar, high fat content), and common oats (*Avena sativa*). We have put the seed through a sterilization process by baking it at 180 F for a period of 2 hours. Dried mealworms will also be placed in all traps. Shrews are active day and night and need insects constantly or they will die. This leads to high death rates among this taxon in trapping efforts (as high as 80% for some species). Careful timing of when the traps are laid and picked up will limit mortalities, as will the additional protein provided by mealworms. However, in multi-night situations shrew death may increase. Shonfield (2013) found that shrew mortality increased as temperatures went below 10 °C (50 °F) and during nights with measurable precipitation. Gannon et al. (2007) suggests not trapping during "extreme" temperature or precipitation events, to avoid negatively impacting target species via direct mortality or harming their physical condition. We avoid trapping during nights with expected precipitation (> 30% chance) and temperatures that drop below 50 °F if possible. If temperatures do drop below 50 °F, we include cotton balls within traps for insulation. We elected to utilize Sherman traps for several reasons. First, it is still the standard method of small mammal trapping, and reliably catches a good diversity of species (USGS 2012; Newsome 2015). A variety of techniques are available, such as mesh live trapping (O'Farrell et al. 1994), and track tube indices (Wiewel et al. 2007). However, no technique is more commonly and reliably used across such a variety of habitats as the Sherman trap (USGS 2012, Newsome 2015).

Traps will be assembled and placed every 5m along the monitoring transect line for 200m, with an additional 5 traps placed every hundred meters within 10 meters of the transect line at the discretion of the trap setting teams (total of 50 traps per site). In 2015 we trapped 33 sites for a total of 46 nights to collect initial inventory data for a total of 2,300 trap nights (# nights X # traps). Based upon this initial effort we determined that the period between 15 August and 21 September was the best time to effectively capture both small mammal abundance and diversity (Figure 6). However, based on additional data from past research conducted at Mormon Island we determined that the period from 1 August to 28 September could be appropriate depending on seasonal weather trends. Initial statistical analysis of the data using ordinary least squares linear regression suggested that as temperature increased 1 degree F that an additional 0.37 ± 0.10 small mammals would be captured per 50 traps ($p = 0.0006$, $R^2 = 0.24$). Therefore, even while trapping during this scheduled time period it may be important to avoid trapping on nights that drop below 50 degrees F to ensure a representative sample. Total captures per night ranged from 0 to 22 out of possible 50 traps, with a mean value of 4.56/50, a lower quartile (25th) value of 1/50, and an upper quartile (75th) value of 7/50. Total species richness detected per night ranged from 0 to 7 species with a mean value of 2.16, a lower quartile of 1 species, and an upper quartile of 3. The 12 core sites surveyed annually are trapped for three nights within a 1-week period. Traps will be set 2-3 hours before sunset and checked within 2 hours of sunrise.

Figure 6. Sample species richness (red) and total captures (blue) of small mammals in 2015 by Julian date (day of year 1-365).

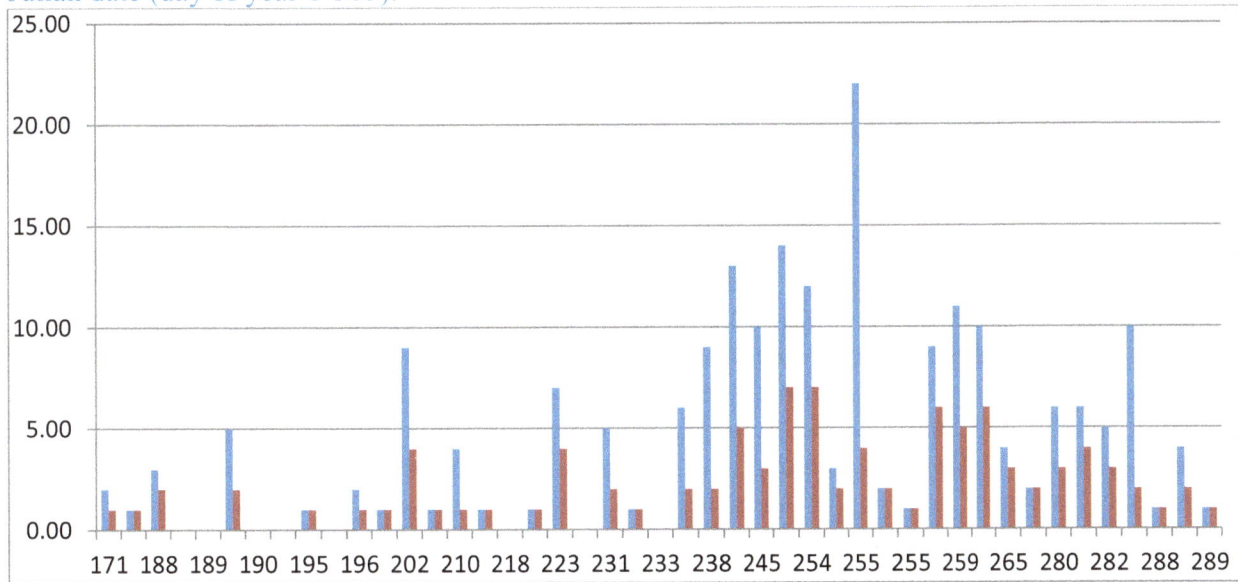

When arriving at a site, locate the t-post marking the beginning of the vegetation transect, looking toward the end of the transect via the specified bearing, use the "right" transect placed 10m to the right-hand side of the of the t-post. This transect can be located by following a bearing perpendicular (90 degree) to the transect bearing and walking 10m to the right. You should be able to locate a tile or capped rebar. Run the transect out via tape measure in the specified transect bearing for 100m and you should be able to locate a second capped rebar. Place a trap every 5 meters on this transect, and then repeat this process following the same bearing from the capped rebar at the 100m mark. Basically, run the tape measure again from the 100m mark in the same bearing to the 200m mark and place 20 more traps (one for every 5m), in addition to 10 across the 200 m stretch at the discretion of the trapping team. These incidental traps can be placed anywhere along the line. It is sometimes best to put all 10 traps in a location that has high traffic, but generally 5 incidental traps on the first 100m and the next 5 on the next 100m gives you the best sampling. Incidental traps should be placed within 10m of the sampling line. The goal of this effort is to examine if these traps perform better when placed on wildlife trails detected by biologists. Each 200m transect will have 50 traps. Best place each "regular" trap within 50cm of the specified 5m mark up or down the transect line (within 50cm of the 5m, 10m, 15m marks, etc.). Try to place the trap in a vegetative gap where small mammals might be moving. Drop 50-75% of the bait within the trap while creating a trail into the trap with the remaining bait. Make sure to test the trap's functionality after assembling it before baiting and placing it. Do your best to make sure the trap at minimum touches the transect line on 1 corner.

Data Management

Data will be recorded in the field on the small mammal datasheet. These sheets should be scanned and emailed to bostrom@cranetrust.org after being entered into the small mammal trapping database. They will then be saved in the Small Mammal Monitoring folder on the Company X drive, located within the Science Program folder. All categories on the datasheet should be filled out for every capture except for morphometric measurements (Total Length, Tail

Length, Hind Foot Length, Ear Length and Weight). These variables are only necessary to record for hard to identify species, or species that are difficult to age and/or sex. Record all morphometric variables in millimeters (mm), except for weight which will be recorded in grams (g). However, recording this data for extreme examples –species either on the high or low side of weight and morphometric measurement ranges- can provide important scientific data and is recommended pending available time. It is also helpful for biologists new to the project to measure all captured individuals as this improves identification accuracy. The "comments" section of the datasheet also does not need to be filled out for every observation. However, any valuable information regarding external parasites, injuries, or morphometric characteristics that do not have their own data category and are important for species identification, for example, the number of toes on the hind foot (4 or 5) or coloration should be recorded in the "comments" section of the datasheet when appropriate. Also include notes regarding trapping such as deceased or unhealthy-looking animals in the "comments" section of the datasheet. Photos of small mammals and small mammal trapping procedures should be saved on in the "Small Mammal Photos" folder located along with Small Mammals Monitoring Database, Small Mammals Datasheet, and Small Mammals Protocol on the X drive in the Small Mammals Monitoring folder. Please label the photos with the species name and date of capture, for example, "Peromyscus_maniculatus_06152015." Trapping procedures, layouts, and activities should be labeled the same way, with the topic followed by the date, for example, "Measurement_Techniques_06152015."

Handling of Small Mammals

To conclude this protocol, I would like to emphasize safety and cleanliness. Please do not handle rodent feces directly. We will periodically spray down all open traps with a light bleach solution. Rodent feces can carry dangerous diseases like Hantavirus. Don't ever put your face or hand directly into a trap without looking. There is always the possibility of a non-target species like a venomous snake (for example, a juvenile *Crotalus viridis* (Prairie Rattle Snake)) being in a trap, even though venomous species are uncommon here. One thing to note is that the Northern Short-tailed Shrew has a venomous bite that will cause you pain for two weeks and is best avoided. To extract a small mammal, place a Ziploc or cloth bag around the entrance of the trap while holding the Sherman trap's door open and gently shake the trap until the species falls into the bag. Utilize the bag to hold the small mammal while you locate the nape behind its head with your thumb, pointer, and middle fingers. Grab as much skin as possible on the nape (back of the head and neck area) without injuring the small mammal to securely hold it and complete the necessary measurements. Shrews do not need to be naped! Weigh them, feed them mealworms, and send them on their way.

Our first priority is the health of the animals. Animals that are lethargic or ill SHOULD NOT be handled. If you have a lethargic shrew it is best to feed it many mealworms and gently rub its body. This will increase the warmth of the animal enough to spur it to eat. If a rodent is lethargic you may release it after spreading some seed down for it. In the case of a severely ill mammal release the animal after noting a physical description, mark the trap with blue tape, place it in the bag the mammal was extracted into and put both of these away. We do not want to spread disease due to trapping. Sanitize hands if possible and sanitize the trap as soon as possible. ALWAYS wash your hands after trapping.

References

Allen, R.P. 1952. The Whooping Crane. National Audubon Society, New York, New York, USA, 246 pp.

Collins, S.L. 2000. Disturbance Frequency and Community Stability in Native Tallgrass Prairie. The American Naturalist 155(3):311.

Horncastle, V.J., E.C. Hellgren, P.M. Mayer, A.C. Ganguli, D.M. Engle, and D.M. Leslie Jr. 2005. Implications of Invasion by *Juniperus virginiana* on Small Mammals in the Southern Great Plains. Journal of Mammalogy 86(6):1144-1155.

Gannon, W. L., and R.S. Sikes. 2007. Guidelines of the American Society of Mammalogists for the use of wild mammals in research. Journal of Mammalogy 88(3):809-823

Johnston, A.N., and R.G. Anthony. 2006. Small-Mammal Microhabitat Associations and Response to Grazing in Oregon. Journal of Wildlife Management 72(8):1736-1746.

Matlack, R.S., D.W. Kaufman, and G.A Kaufman. 2001. Influence of Grazing by Bison and Cattle on Deer Mice in Burned Tallgrass Prairie. American Midland Naturalist 146(2): 361.

Newsome, S. 2015. Small Mammal Mark-Recapture Population Dynamics at Core Research Sites at the Sevilleta National Wildlife Refuge, New Mexico (1989-present). Sevilleta LTER: Long Term Ecological Research, University of New Mexico, Albuquerque, New Mexico, USA.

O'Farrell, M.J., W.A. Clark, F.H. Emmerson, S.M. Juarez, F.R. Kay, T.M. O'Farrel, and T.Y. Goodlett. 1994. Use of a Mesh Live Trap for Small Mammals: Are Results from Sherman Live Traps Deceptive? Journal of Mammalogy 75(3):692-699.

Oswald, C.D. and L.D. Flake. 1994. Bait Formulation and Effectiveness in Live-Trapping Small Mammals in Eastern South Dakota. Proceedings of the South Dakota Academy of Science 73:101-108.

Shonfield, J., R. Do, R.J. Brooks, and A.G. McAdam. 2013. Reducing accidental shrew mortality associated with small-mammal livetrapping I: an inter-and intrastudy analysis. Journal of Mammalogy 94(4):745-753.

US Geological Survey. 2012. Small mammal trapping standard operating procedures. USGS, Western Ecological Research Center, San Francisco Bay Estuary Field Station, Vallejo, CA, USA, 7 pp.

Walkinshaw, L.H. 1949. The Sandhill Cranes. Cranbrook Institute of Science, Bulletin No. 29. Bloomfield Hills, Michigan, USA.

Wiewel, A.S., W.R. Clark, and M.A. Sovada. 2007. Assessing Small Mammal Abundance with Track-Tube Indices and Mark-Recapture Population Estimates. Journal of Mammalogy 88(1):250-260

Appendix 4. Small mammal trapping datasheet

Small Mammal Trapping Datasheet

Transect ID: _____   Observer (s): _____   Date: _____   Temp: _____   Pg. ____ of ____

Meter	Species	Sex (M/F)	Age (A/J)	Reproductive (Scrotal, Pregnant, etc.)	Total lg (mm)	Tail lg (mm)	Hind lg (mm)	Ear lg (mm)	Weight (g)	Comments

Note: It's only necessary to record morphometric measurements for difficult to identify species.

Dichotomous Key for Live Species of Small Rodentia in Central Nebraska

1a. Tail fur narrow for all or majority of tail... **2**
1b. Tail fur wide with hair more dense than entire fleshy portion... **17**

2a. Inconspicuous eyes and ears (ears partially or totally concealed in fur); body stout and stocky....................... **3**
2b. Conspicuous eyes, ears more visible, with a conical elongated snout... **6**

3a. Large front feet with large strong foreclaws, small eyes, 2 grooves on upper incisors, and external cheek pouches.. **Plains Pocket Gopher (*Geomys bursarius*)**
3b. Front feet and claws small.. **4**

4a. Bicolored tail barely longer than the hind feet 12-24 mm; some flesh of the ears visible. Incisors longitudinally grooved. Eyes are small and close to the nose......... **Southern Bog Lemming (*Synaptomys cooperi*)**
4b. Tail longer than the hind feet ≥24 mm; ears concealed in fur. Incisors not longitudinally grooved.................... **5**

5a. Ventral side light tan or buff colored; tail strongly bicolored. 5 pads on the sole of the foot and 6 mammae.....
.... **Prairie Vole (*Microtus ochrogaster*)**
5b. Ventral side gray or white and may be tinged with light brown; tail faintly bicolored. 6 pads on the sole of the hind foot and 8 mammae. .... **Meadow Vole (*Microtus pennsylvanicus*)**

6a. Tail faintly bicolored, scaly, and sparsely haired; pelage has a harsh, coarse appearance. Ears are extremely rounded and sparsely haired.. **Hispid Cotton Rat (*Sigmodon hispidus*)**
6b. Tail bicolored, furred, and smooth; pelage less coarse.. **7**

7a. Tail short (30-60 mm) and approximately 1/3 the length of the body. Typically tail is bicolored with a white tip, but the tip may be black. Back is sandy brown to dark gray with a white or cream belly.....................................
.... **Northern Grasshopper Mouse (*Onychomys leucogaster*)**
7b. Tail longer than 1/3 the length of the body **8**

8a. Head and body is greater than 160 mm in length. Pure white on throat and white with a gray base on the chest and belly. Long thick whiskers with a moderately haired tail that is clearly bicolored.......................................
.... **Eastern Woodrat (*Neotoma floridana*)**
8b. Head and body is less than 160 mm in length **9**

9a. External fur-lined cheek pouches absent.. **10**
9b. External fur-lined cheek pouches present (Heteromyidae)... **15**

10a. Tail length is 110-155mm (or 1.5 times the head and body length), hindfeet are 2.5 times longer than forefeet.. **Meadow Jumping Mouse (*Zapus hudsonius*)**
10b. Tail is shorter than 100mm (less than 1.5 times the body length)... **11**

11a. Head and body is greater than 80 mm in length; lacking grooves on upper incisors............................ **12**
11b. Head and body is less than 80 mm in length; lengthwise grooves on upper incisors present.......................**14**

12a. Has a unique, broad orange lateral line extending from the cheek to the hindquarters. Thin black eyering with a bicolored tail and tufted hairs (9 mm long tuft).. **Brush Mouse (*Peromyscus boylii*)**
12b. Has mostly a uniform brownish back, lacking orange lateral line **13**

Key to Small Mammals – Soricomorpha

1a. Forefeet small;   tail pointed; teeth pigmented **2**
1b. Forefeet much wider than long with long curved nails; tail short, nearly naked, and rounded; teeth lack pigmentation .... **Eastern Mole (*Scalopus aquaticus*)**

2a. Tail almost as long as body, or head and body length combined **3**
2b. Tail distinctly shorter than body **4**

3a. Black tuft of fur at the end of the tail. Pelage is a dark brown. (Tail 33-45mm)..........................
....**Cinereus Shrew (*Sorex cinereus*)**
3b. Tail has a brown tuft of fur at the tip that is not distinct from tail color. Pelage is a light brown. (Tail 30-41mm)... **Prairie Shrew (*Sorex hayendi*)**
May want to leave all mammals that fit #3 to genus (*Sorex* sp.)

4a. Body is significantly large (head and body is 90-114mm) with robust cranium and distinctive short tail (18-30mm). Pelage uniform silver to black with brown tips...
.... **Northern short-tailed Shrew (*Blarina brevicauda*)**
4b. Body is very small (head and body 52-68mm) with short tail (13-23mm). Pelage brown to gray on the back and lighter (almost white) underneath................ **Least Shrew (*Cryptotis parva*)**

Chapter 5: Water Level Monitoring

Project Goals

Our properties are largely composed of lowland tallgrass prairie and wet meadows which are supported by groundwater sub-irrigation (Currier 1989; Henszey et al. 2004; Brinley Buckley et al. 2021). Changes in the groundwater depth can affect the vegetative community (Currier 1989; Henszey et al. 2004). The relationship between surface water and groundwater are not fully understood (Brinley Buckley et al. 2021). Monitoring the groundwater will help us better understand the relationship of the water table to sustaining these systems in a more arid climate than their eastern counterparts. We will also be able to better comprehend the resilience of these systems to periods of drought.

Project Methods

Project History

In 2011 and 2012, Levelogger sensors measuring fluctuations in ground water levels and staff gauges measuring surface water changes were set throughout the Crane Trust property. A total of 14 transducers (6 on Shoemaker Island, 6 on Mormon Island, and 2 on the Dipple property), 1 barologger (on Shoemaker), and 9 staff gauges (6 on Shoemaker and 3 on Mormon) were installed (Table 5; Solinst account #: T817UV). By September 2017, most of the Leveloggers had become unfunctional and were not recording measurements. There was little information available on the original installation and a formal protocol was set up at this time.

Table 5. List of original groundwater monitoring sites and their status as of winter 2017. GPS coordinates and prior metrics before re-installation available on X-drive

Transducers/barologger	Serial #	Functional	Date of Failure	Outcome
Wild Rose HQ Slough (baro)	2005771; 2099024	Y	-	Replaced on 5/9/2019. Continuing.
Wild Rose HQ Slough (trans)	2006051	N-sent to Solinst for assessment	Fall2016-Spring2017	Replaced w/ Dipple upland well logger
WildRose_east_caddis	2016819	Y	-	Continued
WildRose_west_caddis	Unknown	N-logger missing	Spring2013	Will not replace
Type_locality_well	2017485	Y	-	Continued
Type_locality_slough	2017486; 2158427	Y	-	Replaced on 7/12/2022. Continuing.
Mormon_middle_well	2016814	Y	-	Continued
Mormon_transect_well	2017142	N-logger missing	Unknown	Will not replace
Mormon_west_well	2017492	Y	-	Continued
Mormon extreme west well	2006049	Y	-	Continued
WildRose_well	2017412	N-logger missing	Summer2015	Will not replace
WildRose_river	Unknown	N-casing and logger washed away	Fall2012	Will not replace
Dipple north slough	2006374	Y	-	Removed
Dipple upland well	2007038	Y	-	Removed
Staff gauges				
WildRose_headquarters_sg	-	Y	Unknown	Lost but replaced and functioning
WildRose_east_caddis_sg	-	Y	-	Continued
WildRose_west_caddis_sg	-	Y	-	Continued
WildRose_exclosure_sg	-	N	Unknown	Will not replace
WildRose_west_slough_sg	-	Y	-	Continued
WildRose_River_sg	-	N	Unknown	Will not replace
Type_locality_slough_sg	-	Y	-	Continued
Mormon_middle_sg	-	Y	-	Continued
Mormon_west_slough_sg	-	Y-staff gauge needed reattached	Unknown	Repaired and functioning

Project Update 2018

The memory of each Levelogger was cleared after a final download in early winter 2017-2018 and have all been reset to run continuously. The continuous setting on the units ensures groundwater records are not disrupted due to memory fill up, deleting the oldest reading automatically to make space for a new reading to be saved. The two functional Levelogger units were removed from the Dipple property due plans to sell the property. One of these units was used to replace the WildRose HQ Slough transducer and the other was installed at a new well on the Martin's Meadow property to monitor groundwater as restoration commences in the slough there. All other non-functional transducers were not replaced.

A new staff gauge was installed in June 2018 at Martin's Meadow along with the Levelogger well. The WildRose River and WildRose Exclosures' staff gauges were not replaced, and all other staff gauges were repaired at their respective existing locations. A 4'x4' protective enclosure (made from cattle panels) was installed around each Levelogger casing and staff gauge site to prevent damage from grazing cattle and bison. All data up to winter 2017 is available in folders (X-drive >Science Program >GroundWaterMonitoring>(Year) Records).

Table 6. Transducer and barologger locations, elevations, dates of operation, and physical values.

Serial #	Transducers/barologger	Latitude (N)	Longitude (W)	Reference Datum Elevation (top of case) (meters)	Reference Datum Height (cm) above Ground	Unit Length (cm) (Line + 14 cm of transducer)	Date of Re-deployment
2005771	WildRose HQ Slough (baro)	40.79196	98.46103	579.0	179	32	11/14/2017; 5/9/2019
2007038	WildRose HQ Slough (trans)	40.79196	98.46103	579.0	179	141	11/20/2017
2016819	WildRose_east_caddis	40.79570	98.44436	582.0	145	132	11/20/2017
2006374	MartinMeadow slough	40.77227	98.47597	585.0	70	147	6/29/2018
2017485	Type_locality_well	40.80763	98.38424	572.0	125	244	11/27/2017
2017486	Type_locality_slough	40.80779	98.38397	574.6	26	139	11/28/2017; 7/12/2022
2016814	Mormon_middle_well	40.80191	98.40873	577.2	93	182	11/20/2017
2017492	Mormon_west_well	40.79514	98.42656	579.9	68	328	11/20/2017
2006049	Mormon extreme west well	40.78963	98.43555	581.0	81	564	11/20/2017
	Staff gauges			**Staff gauge Zero Point Elevation (bottom of staff gauge) (meters)**			
-	WildRose_HQ_slough_sg	40.79196	98.46103	577			
-	WildRose_east_caddis_sg	40.79570	98.44436	577			
-	WildRose_west_caddis_sg	40.79446	98.44535	578			
-	WildRose_west_slough_sg	40.78439	98.47080	585			
-	Martin_Meadow_Slough_sg	40.77222	98.74593	584			
-	Type_locality_slough_sg	40.80779	98.38397	575			
-	Mormon_middle_sg	40.80151	98.40865	573			
-	Mormon_west_slough_sg	40.79569	98.42706	575			

Figure 7. Groundwater monitoring locations as of 2018 on Mormon and Shoemaker Islands as well as on Martin's Meadows.

Equipment

The Crane Trust currently has 8 transducers, 1 barologger, and 8 staff gauges. The barologger is suspended above a maximum water level and records atmospheric pressure. The transducers are submerged below a minimum water level, recording water pressure that increases as water levels rise. The barologger is used to compensate for the additional pressure from the atmosphere, which is subtracted from the synchronized transducer readings. This compensation yields an accurate water pressure which can then be used to determine the depth of the transducer. Transducers in slough areas monitor surface water level changes, while transducers at well sites monitor subsurface groundwater changes. Staff gauges are used to monitor surface water level changes as well.

Figure 8. Water monitoring units

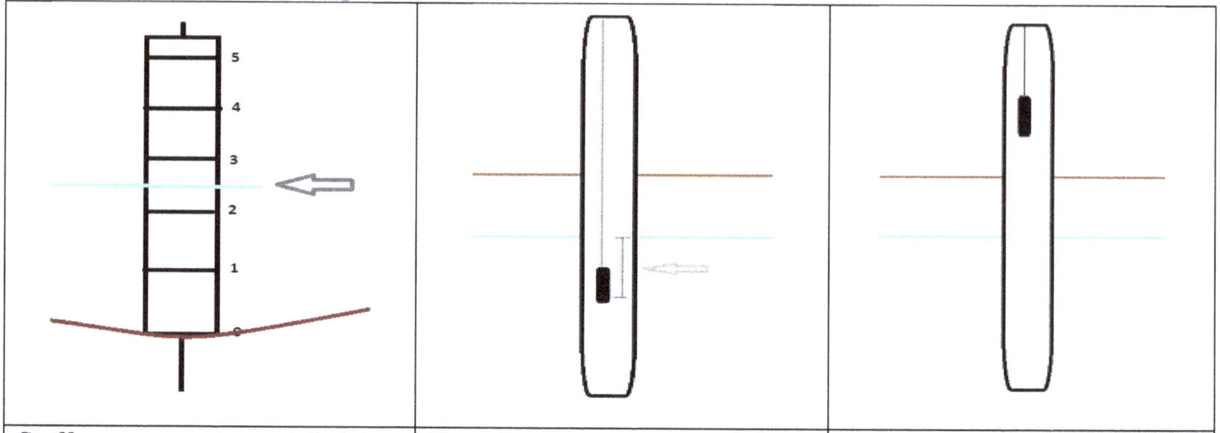

| Staff gauges – post-mounted plates marked with graduated lines which are used to measure the depth (in meters) of surface water above ground level (bottom of slough). Bottom of the slough is at level 0 and the height of the water on the gauge is recorded. | Transducer – sensor suspended below minimum water level in a vented well casing that records temperature and total pressure (kPa) (water + atmospheric) above the sensor zero point, as groundwater raises, the sensor depth increases and a higher pressure is recorded. | Barologger – sensor suspended above maximum water level in a vented well casing that records temperature and atmospheric pressure (kPa) and is used to calibrate all transducer measurements within a 30-mile radius. |

The casings consist of 2" pipes (usually PVC) that are capped off at both ends. Holes were drilled into the bottom of the casing to allow water to enter and exit them as the water level changes. The top cap has one hole in the center for an eye bolt (secured on the topside with a nut) to hang the suspension string from and another offset hole to vent the casing, allowing air pressure inside the casing to equal the air pressure outside of the casing. The casings were dug into the ground and the bottom set below the water table. The Type Locality Slough casing was placed deeper into the slough bed to avoid the risk of damage to the transducer from freezing and ensure transducers remain submerged. Transducers not submerged do not measure water table height, and therefore should remain below an estimated minimum at all times.

The depth of ground water is calculated using the barologger atmospheric pressure (B) reading and the transducers' pressure (L) reading during the same moment in time (Figure 8). The Solinst Levelogger program comes equipped with a Data Compensation Wizard which uses a known factor of increased pressure created by water depth to calculate the submerged depth of the transducer (A) at that moment in time. A known elevation of the top of the casing (reference datum) minus the unit length to zero point (string length + 14 cm) can then be used to calculate the elevation of the groundwater at each well.

The accuracy of the Levelogger water table calculation can be tested at the well sites. A water level meter is used to measure the distance between the top of the casing and the groundwater. The water level meter is a sensor near the end of a measuring tape. When the sensor reaches the groundwater, the tape reel will beep and the distance can be recorded. The recorded distance is subtracted from the reference datum elevation to find the elevation of the groundwater at the well. For slough wells that have a staff gauge, the staff gauge measurement can be added to the elevation at the bottom of the gauge as a second way to test the Levelogger accuracy.

Figure 9. Estimating groundwater depth in relation to surface water elevation.

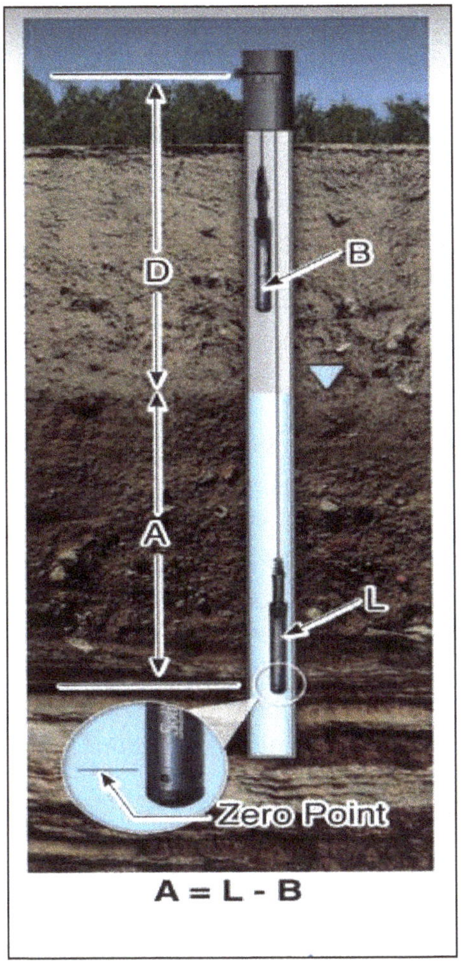

Equations:

Submerged Transduce Depth(A)=

Transducer pressure(L) – Barologger atmospheric pressure(B)

(A=L-B)

Groundwater depth below top of casing(D) =

A – Unit Length

Unit length =

String length from top of casting + 14 cm(top of sensor to zero point)

Or

A+D

Elevation of water table =

Reference datum elevation – Unit length + A

Or

Reference datum elevation – Water level meter reading

Or

Staff gauge reading + Elevation at bottom of staff gauge

Notes: Figure adapted from Solinst (2022).

Levelogger Downloads

The transducers and barologger should be checked and downloaded at least twice per year (Spring and Fall). The loggers are set to run continuously and can store ten years of data, but it is important to make sure well casings and Levelogger units are still functional. Biannual downloads also help to minimize data loss in the event of Levelogger failure or a destroyed well site. Simultaneously, the water level meter reading at each transducer site and each staff gauge should be checked. The standardized datasheet will be used to record the dates of Levelogger downloads, water level meter readings, and staff gauge readings (Appendix 6). Data is downloaded in-situ via the red Dell laptop, or any other laptop computer loaded with the Solinst Levelogger Series 5 Software (available at: https://downloads.solinst.com/solinst-software-firmware-downloads?login_redirect=1). The transducers and barologger use two types of optic port cables that plug into the USB outlet on the laptop. The older transducers use a two-optic cable (shared with the USFWS ES office), and the newer transducers use a large singular-optic cable (owned by the Crane Trust).

Well Water Level Meter

The water level of each Levelogger well should be taken when wells are visited for download. These readings will help set the parameters for data compensation and can be used to check the accuracy of the Levelogger readings. We will use the Solinst Water Level Meter (Model 101) to check the depth of the water from the top of the well casing. Ensure battery is replaced before checking wells! The sensitivity dial should be set to a minimum. It can be calibrated by taking multiple readings at each well, retracting the meter tape each time to ensure contact with water. A clean paper towel should be used to dry off the water level meter probe after each deployment into a well. When sensitivity is minimized, deploy the probe down the well casing slowly until the meter "beeps". Record meter tape measurement from the opening of the casing when the meter "beeps" onto the Groundwater Monitoring Datasheet for each corresponding well.

Staff Gauge

The water level at each staff gauge will also be recorded when Leveloggers are downloaded. Staff gauges are used to measure surface water depth and are mounted to a t-post with the "zero point" on the gauge resting on top of the ground or at the bottom of a slough. Staff gauges are read at the level that the top of the water crosses on the gauge. If there is no water at the staff gauge, record "< 0" on the Groundwater Monitoring Datasheet for the corresponding gauge. It may be necessary to wipe staff gauges off to see graduation lines on the gauge. Structural integrity and condition at each visit for staff gauges should also be evaluated and any repairs made and documented on the Groundwater Monitoring Datasheet at each visit and, which include paint chipping, readability of the staff gauge face, deteriorating or loose staff gauge mounting board, or leaning gauge or t-posts.

Data Management

Downloading Data

Before retrieving data, we will need a computer with the Solinst software, the optic port cables, clean paper towels, clean water, a meter stick and nylon string. The Levelogger units are suspended from a string that attaches to an eye-bolt on each well casing cap. Caution should be taken removing the cap from the casing, gradually twisting the cap off of the casing if it does not come off easily. The string and Levelogger should be slowly pulled out of the casing to avoid damage to the sensor and prevent knotting of the suspension string. A paper towel will be used to dry off and remove surface debris before the sensor cap is unscrewed to ensure sensor lenses remain dry and clean. Excessive mud or dirt may be rinsed with clean water and dried with the sensor cap on. (additional cleaning steps are available in the Levelogger User Guide for corrosive or calcified debris on the sensor). Notes of unit condition and any evident damage to the sensor will be taken and recorded on the datasheet at this time. The condition of the string should be checked for any knotting or cuts in the string length. The knots should be examined at both ends of the string (at the casing cap eye-bolt and at the sensor unit cap) to ensure the string will remain secure. If the string is damaged or the knots appear loose, the string should be replaced after download using the exact length that corresponds with each well listed above so that the depth of the unit remains consistent (See Table 6).

After the Leveloggers are retrieved, cleaned, and examined for damage, we can begin to download the data into the Levelogger program, First, connect the optic dock cable securely into a laptop USB port and open the Levelogger program. Unscrew the Levelogger cap to expose optic lenses (wipe any dirt from the lenses on both the sensor and the dock cable with a soft cloth). Place the unit onto the optic dock (the lenses are aligned by turning the unit until it is locked into the optic dock). In the Levelogger program, select the "Data Control" tab, and select the "Download Data" icon. A window may pop up asking to overwrite existing data, select "NO" and save the data as an .xle file using the following naming format to begin the download:

WellName_yearmonthday.xle (ie. "WestMormonWell_20170825.xle")

The Levelogger program uses .xle files to run data compensations. The data is then exported by selecting the "Data Control" tab and selecting the "Export Data" icon and saved as a .csv file to be used as a database file in Microsoft Excel using the same naming convention as above:
WellName_yearmonthday.csv (ie. "WestMormonWell_20170825.csv"

All downloaded and exported data and corresponding datasheets need to be transferred to the X (public)-drive in the office. The date of download for each Levelogger location will be recorded on the Groundwater Monitoring Datasheet. Once data download is complete and the files saved, the unit is ready to be returned. Carefully thread the cap back onto the Levelogger and secure snuggly, but do not use pliers, as overtightening will break the cap and will likely lead to unit failure. Carefully, replace the unit into the well casing by slowly lowering it with the string and avoiding knotting and replacing the cap securely when finished.

X-drive>Science Program>Groundwatermonitoring>(Year) Records>XLE Data/Excel Data

Data Compensations

Data compensation can be done within the Solinst Program to estimate water levels based upon the barologger data. First, open the Data Control tab, select Open File icon and select all files (including the barologger file) that you want to compensate. Run one well site at a time. Then, switch to Data Compensation tab.

- Step 1
  - Select Levelogger file that should be compensated.
  - Deselect barologger file for this list.
  - Click next.
- Step 2
  - Mark the "Barometric Compensation" adjustment feature.
  - Determine if a Barometric Efficiency Adjustment is necessary (See Section 10.1.3.1 in the User Guide).
  - Click next.
- Step 3
  - Determine if the three parameters need to be selected.
    - If one is selected, all three should be selected to not skew the data.

- Can be run without any of the parameters. Levelogger is programed to produce accurate data in typical situations without need to adjust the parameters.
  - Make sure barologger file is highlighted in the shown list.
  - Click finish.
  - The Levelogger program automatically saves a copy of the compensated data as an .xle file and tags the file name with "Compensated"
  - The same procedure is followed to create a .csv file for Microsoft Excel

All compensated .xle and .csv files should be saved in their respective location on the X-drive:

X-drive>Science Program>Groundwatermonitoring>(Year) Records>XLE Data/CSV data>Comp

References

Brinley Buckley, E.M., A.J. Caven, J.D. Wiese, and M.J. Harner. 2021. Assessing the hydroregime of an archetypal riverine wet meadow in the central Great Plains using time-lapse imagery. Ecosphere 12(11):e03829.

Currier, P.J. 1989. Plant species composition and groundwater levels in a Platte River wet meadow. Proceedings of the North American Prairie Conference 11:19-24.

Henszey, R.J., K. Pfeiffer, and J.R. Keough. 2004. Linking surface- and ground-water levels to riparian grassland species along the Platte River in central Nebraska, USA. Wetlands 24:665-687.

Solinst Canada Ltd. (2022). Solinst Levelogger 5 User Guide. Georgetown, Ontario, Canada, 92 pp.

Appendix 6. Water level monitoring datasheet

Ground Water Monitoring Datasheet	Date -						
Transducers/barologger	Latitude	Longitude	Download .xle (y/n)	Export .csv (y/n)	Water Level Meter Reading (Top of case to water) (cm)	Notes (condition of equipment/equipment needs)	
Wild Rose HQ Slough (baro)	40.79196	98.46103					
Wild Rose HQ Slough (trans)	40.79196	98.46103					
WildRose_east_caddis	40.79570	98.44436					
Martin Meadow Slough	40.77227	98.47597					
Type_locality_well	40.80763	98.38424					
Type_locality_slough	40.80779	98.38397					
Mormon_middle_well	40.80191	98.40873					
Mormon_west_well	40.79514	98.42656					
Mormon extreme west well	40.78963	98.43555					

Staff gauges	Latitude	Longitude	Staff gauge reading
WildRose_HQ_slough_sg	40.79196	98.46103	
WildRose_east_caddis_sg	40.79570	98.44436	
WildRose_west_caddis_sg	40.79446	98.44535	
WildRose_west_slough_sg	40.78439	98.4708	
Martin Meadow Slough_sg	40.77222	98.74593	
Type_locality_slough_sg	40.80779	98.38397	
Mormon_middle_sg	40.80151	98.40865	

Chapter 6: Fish Seining and Slough Condition Monitoring

Project Goals

This program is intended to monitor the condition of select permanent "warm-water slough" wetlands at the Crane Trust on an annual basis. Research indicates that these are important foraging habitats for Whooping Cranes and essential to a variety of migratory waterbirds (Caven et al. 2022). Fish are a key indicator of aquatic ecosystem health and this protocol pairs seining with other metrics of water quality and habitat condition (Hoagstrom et al. 2011). Additional objectives of this program include: tracking the abundance of native state-listed small-bodied fish (e.g., Plains Topminnow – *Fundulus sciadicus*) in relation to exotic-invasive fish species (e.g., Western Mosquitofish – *Gambusia affinis*; Schumann et al. 2015), examining the response of the fish communities to differing riparian management actions (e.g., fire) and grazing regimes (Bison vs. Cattle, variation in stocking rate; Grudzinski et al. 2018), and tracking slough condition and fish community composition (and mortality events) in response to hydrological variation (flood pulses, droughts; Goldowitz and Whiles 1999). This program can also be extended as needed to assess side channels as well as the main channel of the Platte River to answer specific questions related to management (e.g., impact of disking), stream condition (e.g., low flow/fish kill events), and restoration impacts (e.g., woody clearing around slough).

Project Methods

We use a fine mesh seine net to capture fish as research indicates it is an effective method to collect a relatively representative sample of small-bodied fish in lower order streams (Onorato 1998). Additional implements necessary to complete this protocol include a hand net, two 5-gallon buckets, a YSI Pro 1020 water quality meter (Yellow Springs, OH), a turbidity tube (Science First, Yulee, Florida), two metal sieves for benthic invertebrate sampling, two white plastic dish pans for invertebrate sorting, a metric meter stick, and a rubber ball for estimating flow. Field guides (Page and Burr 2011, Tomelleri and Eberle 2011) and internal visual references (Figures 10, 11) should also be brought to sampling sites to confirm fish and invertebrate identifications.

Our YSI unit measures pH, dissolved oxygen, and temperature in real time. This unit should be calibrated per the instruction manual before use weekly following the manufacturer's guidelines (YSI Pro 1020 User Manual). Calibrate the galvanic dissolved oxygen sensor by putting just a bit of water into the temporary storage cup then slightly threading it onto the probe housing. Wait until dissolved oxygen readings stabilize then press enter. Choose a three-point calibration for pH, immerse the sensor into pH buffer solutions with values of 4.0, 7.0, and 10.0, and in each case wait until pH readings stabilize, then press enter. Between slough assessments probes can be stored in deionized water for multiple weeks. However, after this protocol is compete for the year, probes should be removed from the bulkhead, sensor ports should be covered with rubber caps, and probes should be stored properly. The pH sensor should be stored in a 4.0 buffer solution, make sure it is totally immersed or it could mold. The dissolved oxygen sensor membrane should be removed and the sensor itself should be washed thoroughly with deionized water previous to long term storage under a rubber cap. A new sensor membrane should be installed previous to each field season. This will require filling a new membrane cap with "galvanic oxygen sensor electrolyte", making sure to let the solution rest so bubbles disappear before the fluid filled cap is attached onto the galvanic oxygen sensor probe.

Site (e.g., Calving Pasture Slough, "CPS"), date, start and end time, starting and ending GPS locations, all observers names, air temperature, and sky conditions should be recorded near the end of each survey. Each survey will consist of 6-8 seine pulls (6 high or 8 low volume capture events). Seine efforts should attempt to equally target bank and central portions of slough or other waterways. After each seine pull the fish will be dumped from the net into at least two 5-gallon buckets. Teams will then count and identify the fish by pulling them out from the bucket and releasing them back into the slough. Priority sampling areas include two reaches of both Big Slough and Calving Pasture Slough. Each sampling reach is about 300 m of slough length accounting for sinuosity. However, 6-8 seine pulls targeted toward capturing the various types of habitats present generally only traverses about 150 m. Considering wide fluctuations in water levels it is often the case that several areas are too shallow or occasionally deep to sample. Our approach allows us to sample whatever portion we can of each reach. For this reason, it is important to track starting and ending locations per visit to each slough or river channel. Our goal is to sample two spatially distinct reaches of each slough. We additionally included a map and suggested sampling locations for two reaches of the north channel that could be targeted for restoration. These reaches can be surveyed as time allows between August 1st and September 21st. Calving Pasture Slough reach one starts at 40.792208°N, -98.461807°W and reach two starts at 40.793353°, -98.459322°W. Big Slough reach one starts at 40.795308°N, -98.444755°W and reach two starts at 40.796860°N, -98.442336°W.

A number of habitat variables will be recorded at each seine pull location including dissolved oxygen, water temperature, and pH using the YSI 1020 (choose an open column of water so vegetation does not impact results); location of the seine pull relative to the bank (middle or bank); mean elevation rise within 5 m of each bank; the dominant facultative wetland and obligate wetland plants detected; the presence of submerged aquatic plants; mean water depth across 3 equally spaced measurements spanning the width of the slough (at least 0.1 m off the bank); slough wetted width; flow estimated by timing the duration required for a floating ball to travel 1 m (make sure the ball is heavy enough not to be driven by normal winds); and, the percentage of the substrate visually estimated to be sand and gravel or organic matter. Finally, each seine effort will be paired with two separate benthic invertebrate samples. Observers will run two sieves along the substrate of the slough for a distance of less than 1 meter and then search through the collections for macroinvertebrates indicative of water quality including Odonata, Ephemeroptera, and Trichoptera. This is a presence/absence survey and thus should be a relatively rapid assessment. Observers will have white rubber tubs in which to pour benthic samples for sorting, adding just a little water to the sample can often make small inverts easier to find as they tend to rise to the top of the sample.

References
Caven, A.J., A.T. Pearse, D.A. Brandt, M.J. Harner, G.D. Wright, D.M. Baasch, E.M. Brinley Buckley, K.L. Metzger, M.R. Rabbe, and A.E. Lacy. 2022. Whooping crane stay length

in relation to stopover site characteristics. Proceedings of the North American Crane Workshop 15:6-33.

Commonwealth Scientific and Industrial Research Organization (CSIRO). 2020. Ephemeroptera: mayflies. Insects and Their Allies, Commonwealth Scientific and Industrial Research, Canberra, Australia.

Gibb, T.J. 2014. Contemporary insect diagnostics: the art and science of practical entomology. Academic Press, Cambridge, Massachusetts, USA, 345 pp.

Goldowitz, B.S. and M.R. Whiles. 1999. Investigations of Fish, Amphibians and Aquatic: Invertebrate Species Within the Middle Platte River System. Final Report, Platte Watershed Program Cooperative Agreement X99708101. USEPA region VII, Kansas City, Missouri, USA.

Grudzinski, B., C.M. Ruffing, M.D. Daniels, and M. Rawitch. 2018. Bison and cattle grazing impacts on baseflow suspended sediment concentrations within grassland streams. Annals of the American Association of Geographers 108(6):1570-1581.

Hoagstrom, C.W., J.E. Brooks, and S.R. Davenport. 2011. A large-scale conservation perspective considering endemic fishes of the North American plains. Biological Conservation 144:21–34.

Onorato, D.P., R.A. Angus, and K.R. Marion. 1998. Comparison of a small-mesh seine and a backpack electroshocker for evaluating fish populations in a North-Central Alabama stream. North American Journal of Fisheries Management 18:361-373.

Page, L.M., and B.M. Burr. 2011. Peterson field guide to freshwater fishes. 2nd edition. Houghton Mifflin Company, Boston, Massachusetts, USA.

Sánchez-Herrera, M., and J.L. Ware. 2012. Biogeography of dragonflies and damselflies: highly mobile predators. Pages 291- 306 in L. Stevens, editor, Global Advances in Biogeography. IntechOpen Limited, London, United Kingdom, 376 pp.

Schumann, D.A., W.W. Hoback, and K.D. Koupal. 2015. Complex interactions between native and invasive species: investigating the differential displacement of two topminnows native to Nebraska. Aquatic Invasions 10:339–346.

Tomelleri, J.R., and M.E. Eberle. 2011. Fishes of the Central United States. University Press of Kansas, Lawrence, KS, USA, 192 pp.

Figure 10. Visual guide to the fish of the Central Platte River Valley with photos derived from Page and Burr (2011), Tomelleri and Eberle (2011), as well as in-house photography.

Common Carp

Creek Chub, spawning male

River Carpsucker

Creek Chub, non-spawning

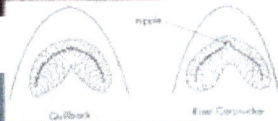

Quillback

Speckled Chub*

White Sucker

Silver Chub

Shorthead Redhorse*

Flathead Chub

Iowa Darter, spawning male

Iowa Darter, spawning male

Johnny Darter

Gizzard Shad

Brook Silverside

Brook Stickleback

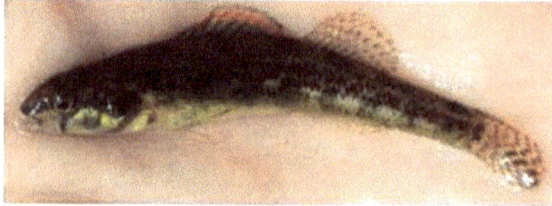

Yellow Perch

Brassy Minnow

Western Silvery Minnow

Plains Minnow

Northern Redbelly Dace, spawning male*

Central Stoneroller, spawning male*

Northern Redbelly Dace, female*

Central Stoneroller, female*

Northern Redbelly Dace, spawning male*

Northern Plains Killifish

Bigmouth Shiner, spawning

Plains Topminnow

Bigmouth Shiner, non-spawning

Western Mosquitofish

Suckermouth Minnow

61

Green Sunfish

Smalllmouth Bass, adult*

White Bass

Smallmouth Bass, juvenile*

Largemouth Bass, adult

Black Crappie, spawning male

Largemouth Bass, juvenile

Black Crappie, female

Common Shiner, spawning male*

Red Shiner, spawning male

Common Shiner, female*

Red Shiner, non-spawning

Fathead Minnow, spawning male

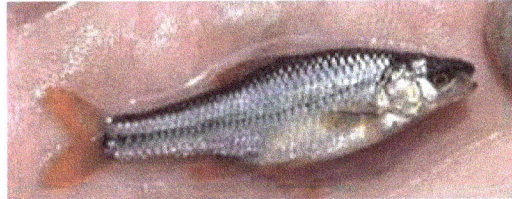

Topeka Shiner, spawning male*

Fathead Minnow, female

Sand Shiner

Emerald Shiner

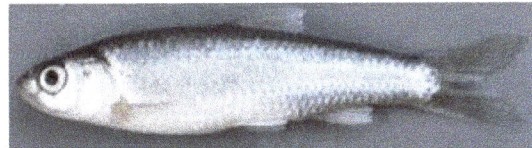

River Shiner

Black Bullhead

Flathead Catfish, juvenile*

Yellow Bullhead*

Flathead Catfish, adult*

Longnose Gar*

Channel Catfish, juvenile

Shortnose Gar

Channel Catfish, female

Mooneye*

Channel Catfish, spawning male

Goldeye

Figure 11. Visual guide to benthic macroinvertebrate indicators of water quality including Odonata, Ephemeroptera, and Trichoptera with photographic and illustrative components. Photographic guide by Emma M. Brinley Buckley. Illustrated guide from Sánchez-Herrera and Ware (2012), Gibb (2014), and CSIRO (2020).

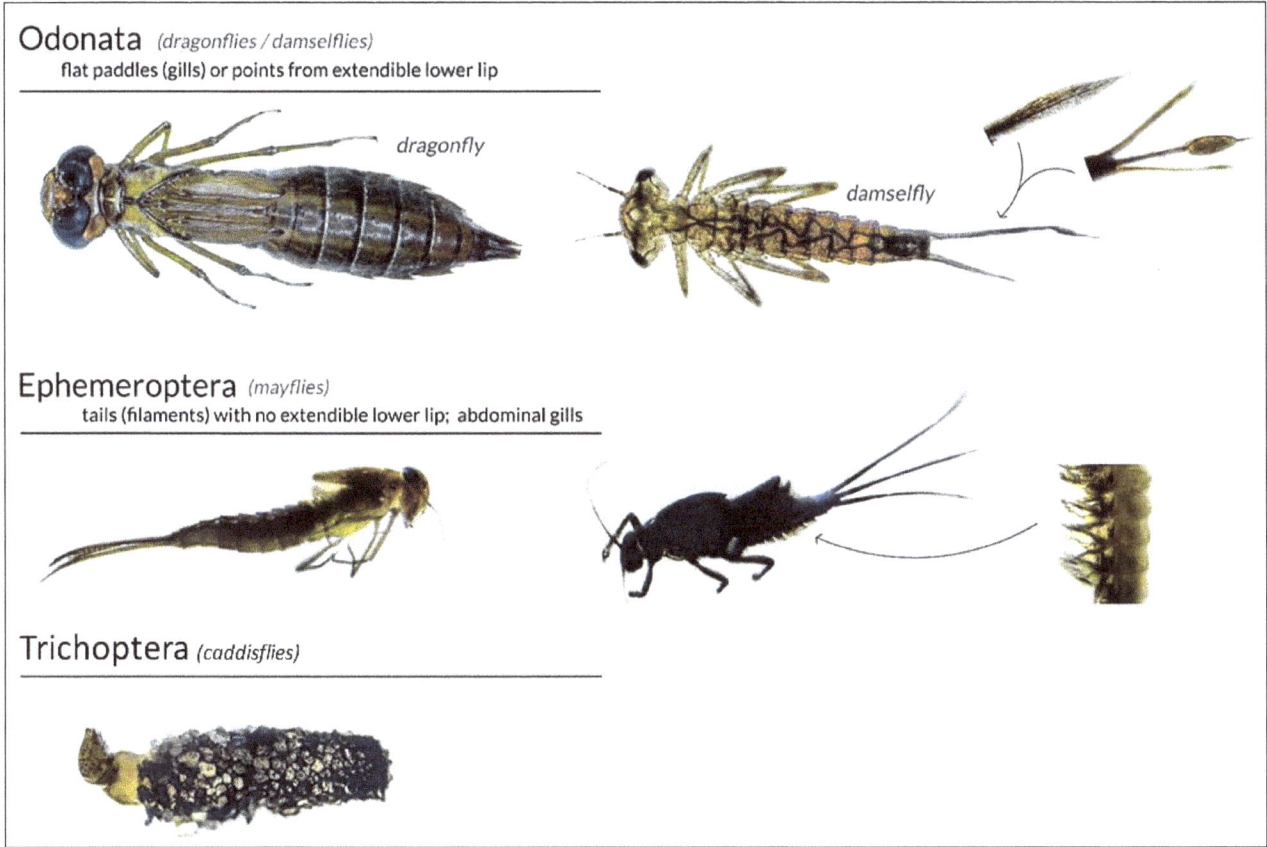

Emma Brinley Buckley (prepared for this document)

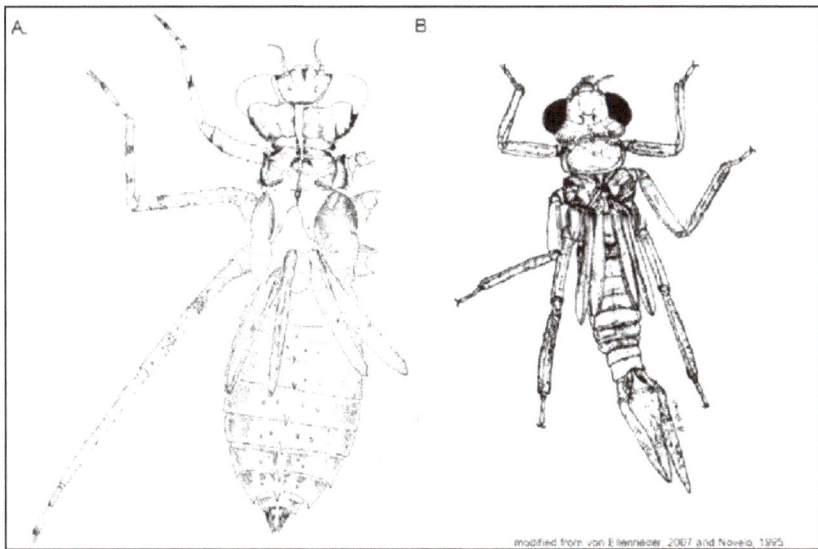

Sánchez-Herrera and Ware (2012): Odonata, Dragonfly (A) and Damselfly (B)

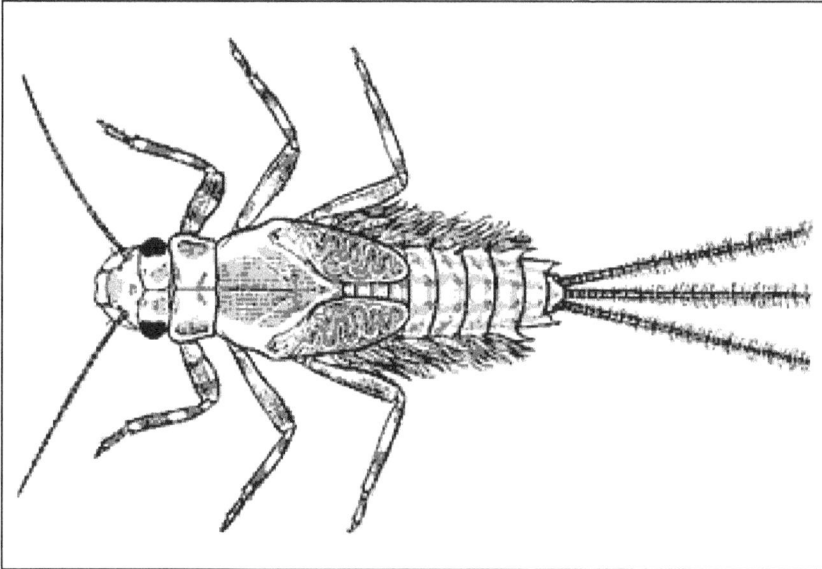

CSIRO (2020): Ephemeroptera, Mayfly.

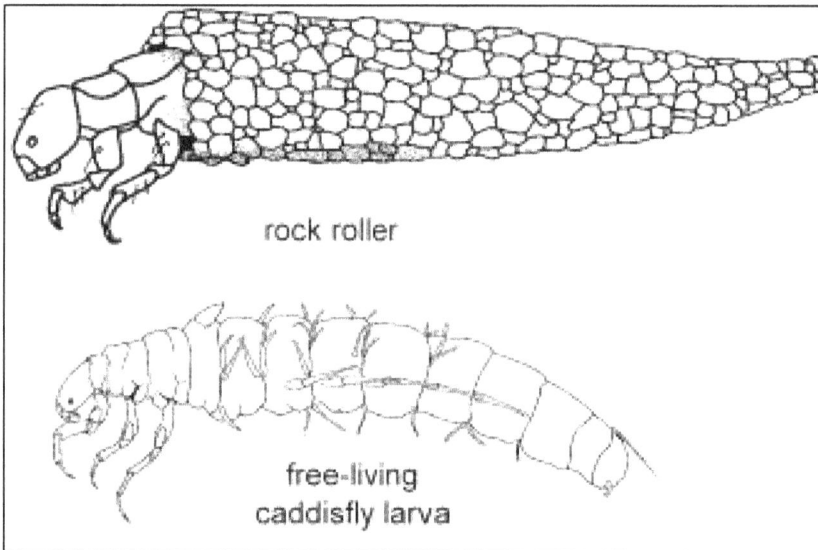

rock roller

free-living
caddisfly larva

Gibb (2014): Trichoptera, encased Caddisfly (top) and free Caddisfly (bottom)

Figure 12. Seining, water quality, and aquatic habitat assessment sites including slough and channel reaches targeted for sampling at Big Slough, Calving Pasture Slough, and on the North Channel of the Platte River.

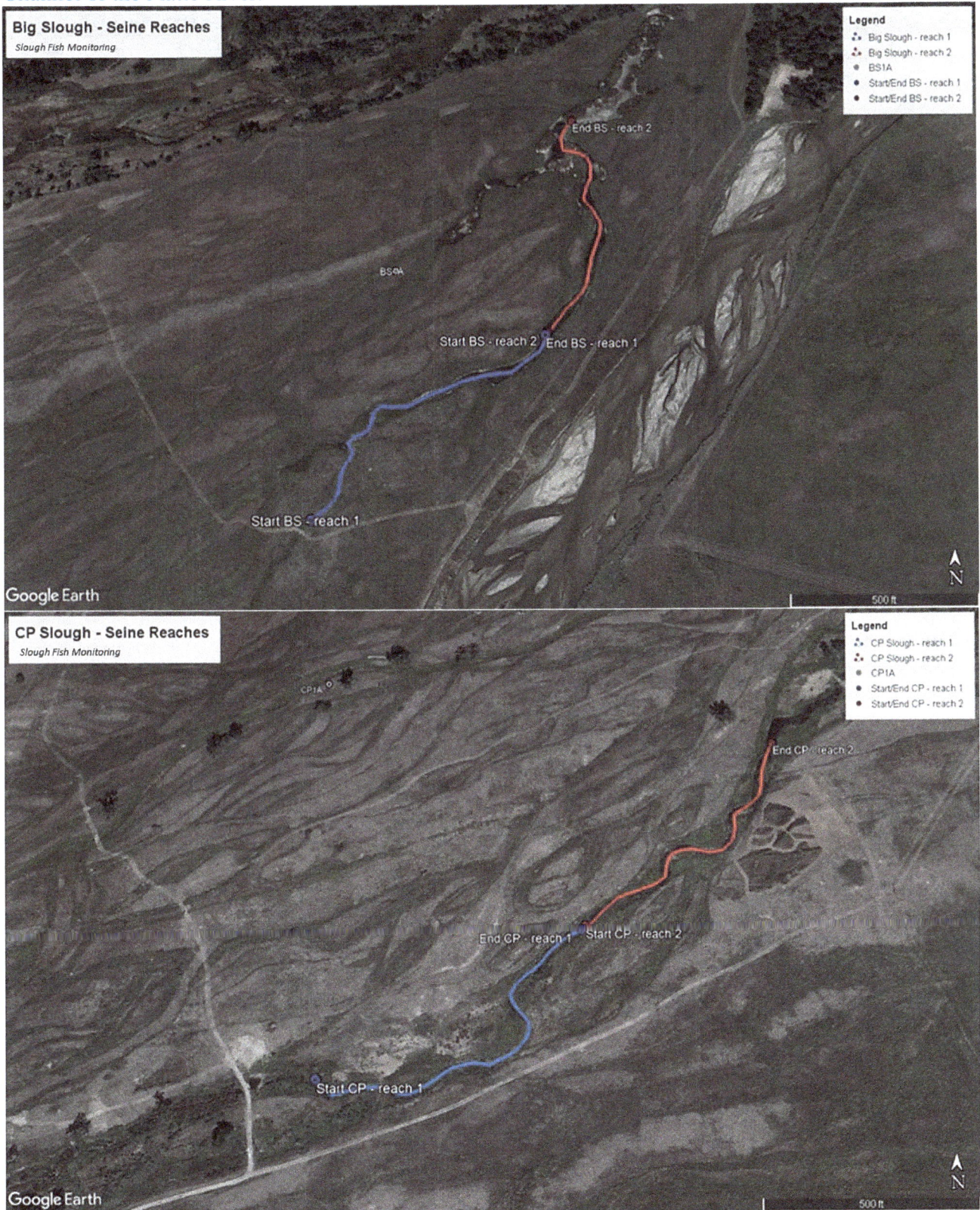

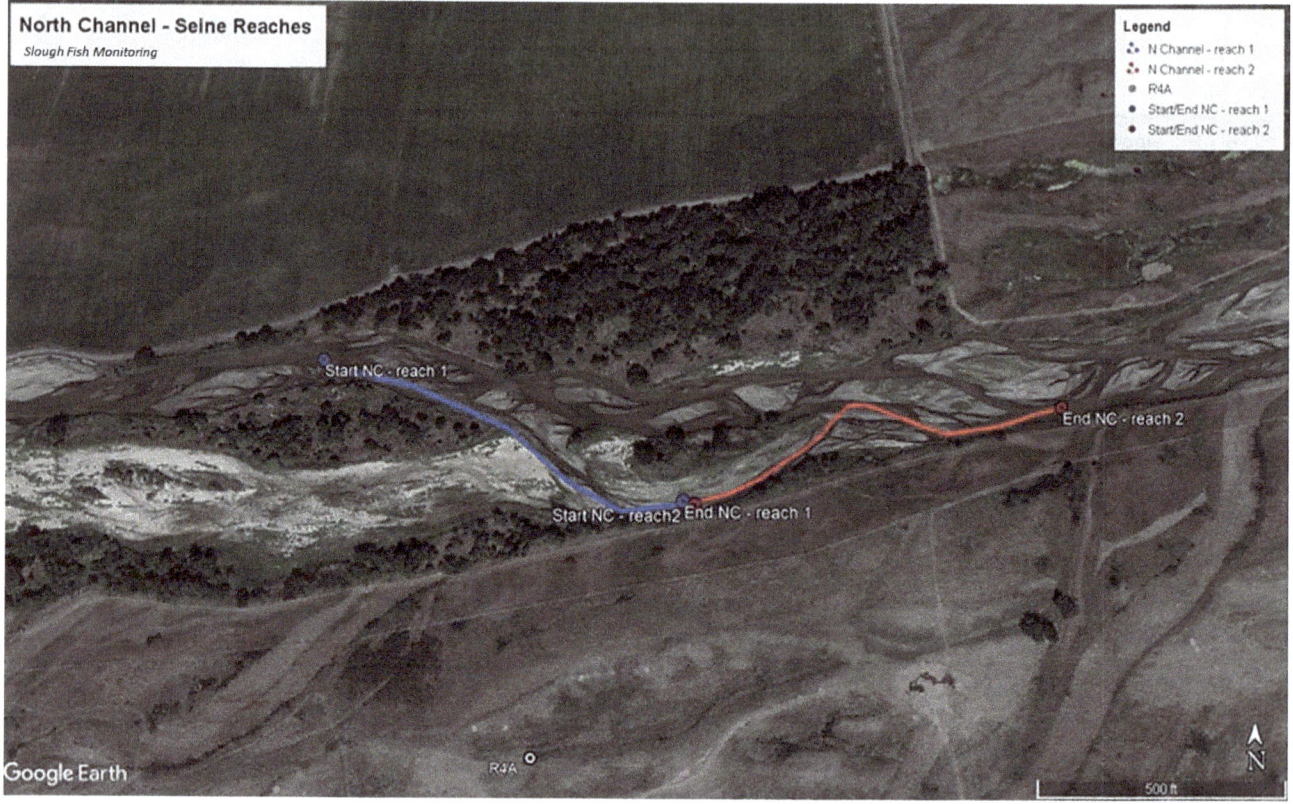

North Channel - Seine Reaches
Slough Fish Monitoring

Legend
N Channel - reach 1
N Channel - reach 2
R4A
Start/End NC - reach 1
Start/End NC - reach 2

Start NC - reach 1
End NC - reach 2
Start NC - reach2   End NC - reach 1
R4A
Google Earth
500 ft
N

Appendix 7. Fish seining and slough condition monitoring datasheet.

Sough-Fish Monitoring Data

Site: _____ Date: _____ St. Time: _____ End Time: _____ Observers: _____ Air Temp (°F): _____ Page: _____ of _____

Sky (clear, cloudy, partly): _____ St. Lat: _____ St. Lon: _____ End Lat: _____ Lon: _____

Run (#)		Depth (cm)		H2O Temp (C)	
Location (bank, middle)		Width (m)		Sand/Gravel (%)	
Bank Slope (%)		Flow (m/s)		Organic (%)	
Dom. FACW Plant		Turbidity (low-high)		Ephemeroptera (y/n)	
Dom. OBL Plant		DO (%)		Trichoptera (y/n)	
Aquatic Plants (y/n)		pH		Odonata (y/n)	
Fish Species				**Number Captured**	

Run (#)		Depth (cm)		H2O Temp (C)	
Location (bank, middle)		Width (m)		Sand/Gravel (%)	
Bank Slope (%)		Flow (m/s)		Organic (%)	
Dom. FACW Plant		Turbidity (low-high)		Ephemeroptera (y/n)	
Dom. OBL Plant		DO (%)		Trichoptera (y/n)	
Aquatic Plants (y/n)		pH		Odonata (y/n)	
Fish Species				**Number Captured**	

Footnotes: Observers- list all participant initials separated by a comma, Turbidity- Low= clear with no reduction in visibility of hand at 15 cm depth, Moderate= murky with reduced visibility of hand at 15 cm depth, High= very murky with visibility of hand at 15 cm depth largely obscured.

69

Appendix 8. Fish species codes and guide

Common Name	Scientific Name	Code
Largemouth Bass	*Micropterus salmonides*	LABA
Smallmouth Bass	*Micropterus dolomieu*	SMBA
White Bass	*Morone chyrsops*	WHBA
Brook Silverside	*Labidesthes sicculus*	BRSI
Brook Stickleback	*Culaea inconstans*	BRST
Black Bullhead	*Ameiurus melas*	BLBU
Yellow Bullhead	*Ameiurus natalis*	YEBU
River Carpsucker	*Carpiodes carpio*	RICA
Channel Catfish	*Ictalurus punctatus*	CHCA
Flathead Catfish	*Pylodictis olivaris*	FLCA
Central Stoneroller	*Campostoma anomalum*	CEST
Creek Chub	*Semotilus atromaculatus*	CRCH
Flathead Chub	*Platygobio gracilis*	FLCH
Silver Chub	*Macrhybopsis storeriana*	SICH
Speckled Chub	*Macrhybopsis aestivalis*	SPCH
Common Carp	*Cyprinus carpio*	COCA
Black Crappie	*Pomoxis nigromaculatus*	BLCR
Longnose Dace	*Rhinichthys cataractae*	LODA
Northern Redbelly Dace	*Chrosomus eos*	NRDA
Iowa Darter	*Etheostoma exile*	IODA
Johnny Darter	*Etheostoma nigrum*	JODA
Longnose Gar	*Lepisosteus osseus*	LOGA
Shortnose Gar	*Lepisosteus platostomus*	SHGA
Gizzard Shad	*Dorosoma cepedianum*	GISH
Goldeye	*Hiodon alosoïdes*	GOLD
Northern Plains Killifish	*Fundulus kansae*	NPKI
Western Silvery Minnow or Plains Minnow	*Hybognathus argyritis* or *H. placitus* (indistinguishable)	WSPM
Brassy Minnow	*Hybognathus hankinsoni*	BRMI
Fathead Minnow	*Pimephales promelas*	FAMI
Suckermouth Minnow	*Phenacobius mirabilis*	SUMI
Mooneye	*Hiodon tergisus*	MOON

Common Name	Scientific Name	Code
Western Mosquitofish	*Gambusia affinis*	WEMO
Yellow Perch	*Perca flavescens*	YEPE
Quillback	*Carpiodes cyprinus*	QUIL
Bigmouth Shiner	*Notropis dorsalis*	BISH
Common Shiner	*Luxilus cornutus*	COSH
Emerald Shiner	*Notropis atherinoides*	EMSH
Red Shiner	*Cyprinella lutrensis*	RESH
River Shiner	*Notropis blennius*	RISH
Sand Shiner	*Notropis stramineus*	SASH
Topeka Shiner	*Notropis topeka*	TOSH
Shorthead Redhorse	*Moxostoma macrolepidotum*	SHRE
White Sucker	*Catostomus commersonii*	WHSU
Green Sunfish	*Lepomis cyanellus*	GRSU
Plains Topminnow	*Fundulus sciadicus*	PLTO

Chapter 7: Butterfly Species of Concern Monitoring

Project Goals

The Regal Fritillary (*Speyeria idalia*) and Monarch Butterfly (*Danaus plexippus*) have both seen recent and rather precipitous declines in their populations (Selby 2007; Brower et al. 2012; USFWS 2015, 2020; Swengel and Swengel 2016). Whether recent declines are a result of continuing habitat loss, or have another primary etiology, they are certainly part of a global decline of Lepidoptera (Dover et al. 2011). Although neither species is federally listed, the Regal Fritillary is listed as endangered, threatened, or a species of concern in 10 states, and it is currently under review for federal listing under the Endangered Species Act (Selby 2007; USFWS 2015). The USFWS (2020) recently concluded that listing the Monarch as an endangered species was warranted but precluded by higher priority departmental actions. Mike Fritz, the former Zoologist at the Nebraska Natural Heritage Program within the Nebraska Game and Parks Commission (NGPC) informed Crane Trust staff in 2015 that the state of Nebraska was beginning to monitor both Monarch and Regal Fritillary populations and that it was possible that these two species could be listed as federally threatened or endangered in the near future (pers. com. 04/05/2015). Therefore, the Crane Trust followed suit and began monitoring Regal Fritillary and Monarch butterflies in 2015 as an effort to contribute regional knowledge to efforts aimed at listing the Regal Fritillary as a federally endangered species, as well as supporting efforts by the NGPC (Caven and King 2015). This endeavor allows us to better understand the impacts of land management on a broad swath of the endemic biota present at the Crane Trust and share that knowledge regionally and beyond. In 2017 we produced a publication in the Journal of Insect Conservation describing our initial findings regarding Regal Fritillary habitat and the impact of land management upon them (Caven et. al 2017). We plan to continue this work indefinitely providing long-term data, clarifying initial findings, and additionally producing a similar publication relating land management and habitat variables to Monarch butterfly abundance.

The Regal Fritillary (*Speyeria idalia* Drury) population has declined by 75-95% since 1990 (Swengel and Swengel 2016). Consequently, for the past two decades *S. idalia* has been listed in many states as a species of conservation concern and is currently a candidate for the federal endangered species list (Selby 2007, USFWS 2015). Investigations into both the characteristics of prairies where Regal Fritillaries make their homes and population trends in these areas are needed if we hope to aid this species to a stable (non-declining) state. Throughout their range these butterflies are found in isolated pockets (Davis et al. 2007, Selby 2007, Caven et al. 2017). In the western extent of their range, these pockets generally become more isolated as patches of appropriate tallgrass prairie habitat become smaller and tied to comparatively mesic lowlands that accumulate just enough moisture to maintain such a community. Over 97% of the tallgrass prairie in Nebraska is gone as a result of development (predominantly agricultural), this figure is even more stark within the eastern third of Nebraska (99%), while the isolated patches of tallgrass prairie further west within river valleys and other lowlands remain somewhat more intact (Noss et al. 1995; Ratcliffe and Hammond 2002). Research indicates that Regal Fritillaries need relatively large, connected tracts of relict prairie including violet species (*Viola* spp.), well drained soils, facultative upland tallgrass prairie species (in particular Big Bluestem, *Andropogon gerardii*), a lack of habitat fragmentation, and moderate management regimes that allow thatch accumulation without allowing significant shrub encroachment (Caven et al. 2017). Regals are sensitive to frequent fire and heavy grazing, and we hope to better understand this given long-term data and varied management strategies (Swengel 1996; Swengel et al. 2011; Moranz et al. 2014, Pierson et al. 2019).

Brower et al. (2012) documented a statistically significant downward trend in the area of wintering Monarch Butterflies in southcentral Mexico from approximately 11 ha of high elevation pine-fir habitat occupied by wintering Monarchs in 1994 to approximately 5 ha in 2011. Brower et al. (2002) demonstrated a 44% decrease in high quality forest and more than a 4-fold decrease in the size of the largest tract of intact forest within key Monarch wintering habitats in Mexico from 1971 to 1999. Degradation of wintering habitat is a key factor negatively impacting Monarch populations. However, habitat loss within the Midwest and Great Plains of the United States and Canada may be playing an even larger role than deforestation in the precipitous population declines (Bowman et al. 2012; Pleasants et al. 2013; Flockhart et al. 2015). Monarch populations are being negatively impacted on their breeding grounds in North America by continued land development as well as the loss of milkweeds from within and on the edges of agricultural fields as a result of glyphosate resistant "Roundup Ready" crops, which eliminate virtually all "weeds" from farm fields (Pleasants et al. 2013; Flockhart et al. 2015). The Crane Trust has implemented land management strategies aimed at promoting native milkweed species (*Asclepias* spp.) within and on the edges of our prairies. This research program allows us to track Monarch use per pasture and serve as a station to monitoring the Monarch migration through central Nebraska into the future.

Project Methods

We counted butterflies using linear walking transects adapted from the methods of Swengel (1996) and Pollard (1977). Plots were examined beforehand to ensure they had appropriate floral resources, in terms of currently flowering plants, for Regals and Monarchs (Nagel et al. 1991; Huebschman 1998; Helzer 2012; Davis et al. 2007; Selby 2007). If significant forb components were not flowering, the survey was delayed until a more appropriate day. During plot visits, butterfly surveys are conducted by two research personnel; the observer spots butterfly species of concern, while the recorder utilizes a GPS and a compass to navigate the monitoring transect, record data, and aid in the detection of butterflies. We count "butterflies observed ahead and to the sides to the limit at which a species can be identified with binoculars" (Swengel 1996). Detections are recorded as within 10 meters of the transect or outside of this area. The recorder should note that only Regal Fritillaries (REFR) species are to be sexed within 10 m of the transect line. The male has a lower line of orange spots on the hind wing, while females have two lines of white spots. Mapping the sex ratio through time may prove to be helpful in the future. Monarchs are not to be sexed since male and female morphological differences are slight, and accuracy may be compromised at a distance. Monarchs and Regals were incidentally recorded on the walk to and from biological monitoring plots utilizing GPS as well. All sightings within 200 meters of the start of a monitoring transect and their corresponding GPS locations should be included as incidental detections on the BSOC datasheet. All other incidentals should be recorded in the BSOC Incidental database. Surveys last 15 minutes, but can be extended if absolutely necessary to accommodate the presence of several butterflies to ensure proper documentation and thorough counts. Surveys are only conducted during favorable weather conditions (sunny, wind under 10mph) between the late morning (10:00am) and the mid-afternoon (4:00pm). All plots are visited at least three times during the Regals' active time period, from June 15th to September 15th. It is advised to visit each plot twice between June 15th to August 1st, to capture male emergence and then the peak Regal activity. The third survey at each plot is recommended to be conducted between August 25th and September 15th, to capture peak female Regal activity. This is based on the timing of Regal activity demonstrated from previous work conducted in the region (Helzer and Jasnowski 2011).

Data Management

For a description of the data collected during BSOC surveys please see Table 1. The BSOC Database, BSOC Datasheet, and places to offload incidental BSOC data not from near or on designated monitoring sites (BSOC Incidental Database and BSOC GPS Data (GPX, KML, or GDB)), as well as a place to offload scanned datasheets and quality BSOC photos, is located on the X-drive under Science Program > Lepidoptera > Butterfly Species of Concern (Regals and Monarchs). Also housed in this location is data from Monarch tagging efforts that will also be a focus of BSOC monitoring in the fall. The BSOC datasheet is to be used only to record survey data. The recorder should note that the BSOC datasheet has a column for incidental sightings. This column is for incidental sightings while walking to/from/between transect sites on a survey day, if the sighting is within 200M of the transect line. If more than one incidental GPS point is taken for one site, the point closest to the transect will be recorded in the "incidental" column, while the other points will be noted in the comments. The BSOC Incidental database is used to record REFR and MOBU sightings >200m away from designated BSOC monitoring plots that are encountered outside of survey periods. This helps ensure an up-to-date estimate of the distribution of both Regals and Monarchs across unmonitored Crane Trust lands.

Table 7. Variable descriptions for butterfly species of concern database and field datasheet

Variable	Variable Description
Site	Code for monitoring plot name
Temp	Temp in degrees F
Wind Speed	Wind Speed in mph
REFR Male 10m	Count of Male Regal Fritillaries within 10m of walking transect
REFR Fem. 10m	Count of Female Regal Fritillaries within 10m
REFR NS 10m	Count of Not Sexed Regal Fritillaries within 10m of walking transect
REFR Out	Count of Regal Fritillaries outside of 10m of walking transect
MOBU 10m	Count of Monarchs within 10m of walking transect
MOBU Out	Count of Monarchs outside of 10m of walking transect
MOBU VP1 (Assoc. Plants)	Monarch use of vascular plants species (most used)
MOBU VP2+ (Assoc. Plants)	Monarch use of vascular plants species (2nd most used and all subsequent plants used separated with a comma)
REFR VP1 (Assoc. Plants)	Regal Fritillary use of vascular plants species (most used)
REFR VP2+ (Assoc. Plants)	Regal Fritillary use of vascular plants species (2nd most used and all subsequent plants used separated with a comma)
INC MOBU n	Incidental count of Monarchs within approximately 200m of the monitoring transect detected off of survey route and/or outside of time period.
INC MOBU Lat. (GPS N, W)	GPS (WGS 84) Latitude from incidental Monarch count (If multiple GPS points with multiple MOBU counts within 200m of monitoring site, average the locations)
INC MOBU Lon. (GPS N, W)	GPS (WGS 84) Longitude from incidental Monarch count (If multiple GPS points with multiple MOBU counts within 200m of monitoring site, average the locations)
INC REFR n	Incidental count of Regal Fritillaries within approximately 200m of the monitoring transect detected off of survey route and/or outside of time period.
INC REFR Lat. (GPS N, W)	GPS (WGS 84) Latitude from incidental Regal Fritillary count (If multiple GPS points with multiple REFR counts within 200m of monitoring site, average the locations)
INC REFR Lon. (GPS N, W)	GPS (WGS 84) Longitude from incidental Regal Fritillary count (If multiple GPS points with multiple REFR counts within 200m of monitoring site, average the locations)
Comments	Observations of other butterflies and pollinators taken during surveys.
*Butterflies detected outside of designated Butterfly Species of Concern Survey Sites should be recorded within the BSOC Incidental Detections Database present on the X (Public)-Drive	

Figure 13. Monarch butterfly identification. Left, Monarch (top) compared to Viceroy (bottom). Right, male (top) as compared to a female (bottom; for educational purposes, sex differentiation will not be made regarding Monarchs during this project).

Notes: Left figure adapted from Journey North Webpage (2017) and Right figure adapted from "Butterfly Garden.net."

Figure 14. Regal Fritillary female (left) as compared to male (right). Regarding hind wing, females have two rows of white to cream colored spots. Males have interior row of white spots, with an exterior row of orange spots on the hindwing.

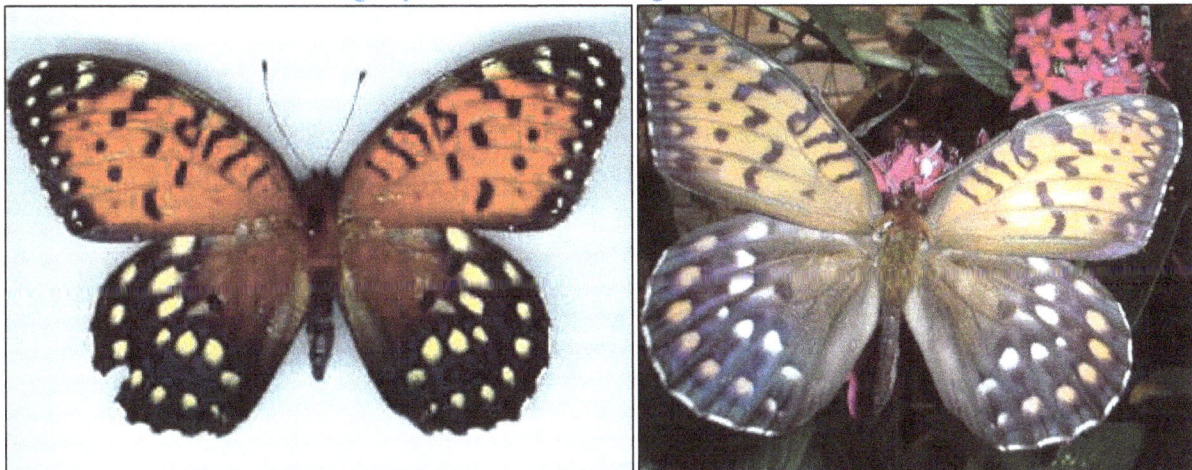

Notes: Regal Fritillary Female (left), image from Great Pains Nature Center. Regal Fritillary Male (right), image from Iowa State University.

Monarch Tagging

Monarch tagging can also be conducted to help cooperate in Monarch monitoring and conservation at a continental scale. The purchase of Monarch tags is through Monarch Watch; tagging can begin in August and goes through the end of their migration (early October). Tags are then reported back via an emailed excel sheet to tag@ku.edu and any leftover tags can be mailed by post back to the organization. The up-to-date protocol for tagging will be included via mail along with the tags. We recommend trying to space out tags across the migration to work alongside their goal of determining which Monarchs, and from where, arrive at the wintering grounds. Tagging may also give us a good idea of the proportion of Monarchs that make it from the Crane Trust to the wintering grounds in Mexico. Recovered tags are reported by the following summer.

References

Brower, L.P., O.R. Taylor, E.H. Williams, D.A. Slayback, R.R. Zubieta, and M.I. Ramirez. 2012. Decline of monarch butterflies overwintering in Mexico: is the migratory phenomenon at risk? Insect Conservation and Diversity 5(2):95-100.

Caven, A.J., K.C. King, J.D. Wiese, and E.M.B. Buckley. 2017. A descriptive analysis of Regal Fritillary (*Speyeria idalia*) habitat utilizing biological monitoring data along the big bend of the Platte River, NE. Journal of Insect Conservation 21:183–205.

Caven, A.J. and K.C. King. 2015. Crane Trust Regal Fritillary Status Review Submission. Report submitted to US Fish and Wildlife Service Region 6.

Dover, J.W., M.S. Warren, and T.G. Shreeve. 2011. 2010 and beyond for Lepidoptera. Journal of Insect Conservation 15:1-3.

Davis, J.D., D.M. Debinski, and B.J. Danielson. 2007. Local and landscape effects on the butterfly community in fragmented Midwest USA prairie habitats. Landscape Ecology 22:1341–1354

Flockhart, D.T., J.B. Pichancourt, D.R. Norris, and T.G. Martin. 2015. Unravelling the annual cycle in a migratory animal: breeding- season habitat loss drives population declines of monarch butterflies. Journal of Animal Ecology 84(1):155-165.

Fritz, M. 2015. Personal Communication. Nebraska Game and Parks Commission, Lincoln, Nebraska, USA.

Helzer, C., and M. Jasnowski. 2011. Results from 2011 Regal Fritillary Butterfly Surveys in The Nature Conservancy's Platte River Prairies, Nebraska. The Nature Conservancy, The Prairie Ecologist blog, Aurora, Nebraska, USA. https://theprairieecologist.files.wordpress.com/2011/12/2011regalbflysummary_platte1.pdf

Helzer, C. 2012. Regal Fritillary butterflies in burned and grazed prairie. The Nature Conservancy, The Prairie Ecologist blog, Aurora, Nebraska, USA. https://prairieecologist.com/2012/11/13/regal-fritillary-butterflies-in-burned-and-grazed-prairie/

Huebschman, J.J. 1998. The relationship between nectar sources and Regal Fritillary (*Speyeria idalia* Drury) butterfly populations. Thesis. University of Nebraska at Omaha, Omaha, Nebraska, USA, 75 pp.

Moranz, R.A., S.D. Fuhlendorf, and D.M. Engle. 2014. Making sense of a prairie butterfly paradox: the effects of grazing, time since fire, and sampling period on regal fritillary abundance. Biological Conservation 173:32-41.

Nagel, H.G., T. Nightengale, and N. Dankert. 1991. Regal fritillary butterfly population estimation and natural history on Rowe Sanctuary, Nebraska. Prairie Naturalist 23:145–152.

Noss, R.F., E.T. LaRoe, and J.M. Scott. 1995. Endangered ecosystems of the United States: a preliminary assessment of loss and degradation. U.S. Department of the Interior, National Biological Service, Washington, D.C., USA, vol. 28.

Pierson, A., A.J. Caven, and C. Wagner. 2019. 40 years of Regal Fritillary data at Audubon's Rowe Sanctuary. Pages 10-11 in A.J. Caven, J. Malzahn, T. Franti, and E.M. Brinley Buckley, editors, Proceedings of the Thirteenth Platte River Basin Ecosystem Symposium, 5-6 June 2018, Wood River, NE, USA. Nebraska Water Center, University of Nebraska-Lincoln, Lincoln, NE, USA, 30 pp.

Pleasants, J.M., and K.S. Oberhauser. 2013. Milkweed loss in agricultural fields because of herbicide use: effect on the monarch butterfly population. Insect Conservation and Diversity 6(2):135-144.

Pollard, E. 1977. A method for assessing changes in the abundance of butterflies. Biological Conservation 12(2):115-134.

Ratcliffe, B.C., and P.C. Hammond. 2002. Insects and the native vegetation of Nebraska. Transactions of the Nebraska Academy of Sciences 28:29-47.

Selby, G. 2007. Regal Fritillary (*Speyeria idalia* Drury): a technical conservation assessment. [Online]. USDA Forest Service, Rocky Mountain Region, Species Conservation Program, Lakewood, Colorado, USA, 53 pp.

Swengel, A.B. 1996. Effects of fire and hay management on abundance of prairie butterflies. Biological Conservation 76(1):73-85.

Swengel, S.R., D. Schlicht, F. Olsen, and A.B. Swengel. 2011. Declines of prairie butterflies in the midwestern USA. Journal of Insect Conservation 15:327-339.

Swengel, S.R., and A.B. Swengel. 2016. Status and Trend of Regal Fritillary (*Speyeria idalia*) (Lepidoptera: Nymphalidae) in the 4th of July Butterfly Count Program in 1977–2014. Scientifica (2016):2572056.

[USFWS] US Fish and Wildlife Service. 2015. 90-Day Findings on 25 Petitions. Department of the Interior, 50 CFR Part 17, Endangered and Threatened Wildlife and Plants. Federal Register 80(181):56423-56432.

[USFWS] US Fish and Wildlife Service. 2020. 12-Month Finding for the Monarch Butterfly. Department of the Interior, 50 CFR Part 17, Endangered and Threatened Wildlife and Plants. Federal Register 85(243):81813-81822.

Crane Trust Butterflies of Concern Datasheet   Date: _____

Observer: _____ Recorder: _____ Temp: _____ Wind Speed _____

Mon Plot: _____ Start Time: _____ End Time: _____

Sp.	In 10 m			Out 10 m	Assoc. Plants	INC n	GPS (N,W)
MOBU							
REFR	Male	Fem.	NS				

Comments: _____

Mon Plot: _____ Start Time: _____ End Time: _____

Sp.	In 10 m			Out 10 m	Assoc. Plants	INC n	GPS (N,W)
MOBU							
REFR	Male	Fem.	NS				

Comments: _____

Mon Plot: _____ Start Time: _____ End Time: _____

Sp.	In 10 m			Out 10 m	Assoc. Plants	INC n	GPS (N,W)
MOBU							
REFR	Male	Fem.	NS				

Comments: _____

Chapter 8: Herpetofauna (Anuran) Monitoring

Project Goals

Due to the recent comprehensive herpetological inventory by Geluso and Harner (2013), there is little need for a full-scale, intensive trapping effort for years to come. Geluso and Harner (2013) utilized pitfall and funnel traps in drift fence arrays to survey Mormon and Shoemaker Islands; they captured 15 total species of herpetofauna, which is 5 more species than Jones et al. (1981) found on Mormon Island. In addition, McLean et al. (2015) detected a Cope's Grey Treefrog during the summer of 2014 on the Crane Trust's Shoemaker property utilizing vocalization surveys. Additionally, in 2018, via this monitoring program we detected a Plains Spadefoot Toad on Shoemaker Island (Wild Rose BS1; Table 9, 10). Therefore, a total of 17 herpetofauna species have been documented on the Crane Trust's lands. B. Ostrom and A. Caven recognized a Blanchard's Cricket Frog (*Acris crepitans blanchardi*) on a SM4 recording from the Central Platte River Valley in 2021, but this species has not yet been positively identified in the field at Crane Trust properties. Following the inventory work of Geluso and Harner (2013), Crane Trust research staff has focused on better describing the habitat and behaviour of species in grassland ecosystems (Wiese et al. 2016a; Caven et al. 2017; Wiese and Caven 2017) as well as mortality risks to herpetofauna communities (Harner et al. 2011, 2013; Wiese et al. 2016a, 2016b, 2017a, 2017b; Schultz and Caven 2021).

Our goal now is to continue to monitor the herpetofaunal response to various management regimes, but in a long-term, low-impact manner. We want to detect broad general changes in species abundance and distribution on the Crane Trust's property over time. Anuran species, including Boreal Chorus Frogs and Plains Leopard Frogs, are certainly part of the migratory Whooping Crane's diet (Allen 1952; Geluso et al. 2013; Caven et al. 2021). These species are likely very important food sources in the freshwater stopovers that Whooping Cranes use during their long migration from coastal Texas to northern Canada. As Whooping Crane conservation is our mission, it will be important to monitor herpetofauna populations for changes over time. Although pitfall and funnel trapping are good methods of sampling and even monitoring in some contexts, they can be labor intensive and can potentially result in high levels of mortality for target and non-target species, if they are not checked frequently. Additionally, in mesic prairies, too many small mammal species are incidentally trapped and drowned. Simply, utilizing pitfall and funnel traps can result in a relatively time intensive and higher impact study when compared with other available survey and monitoring methods that also meet our basic objectives.

Project Methods

We are currently investigating low-intensity methods that allow us to track rather gross changes in responses to management actions. Our overall goal is to achieve detectability, while having a low impact on wildlife, and a relatively minor time commitment. We think the success of the Cope's Grey Treefrog survey work completed by McLean et al. (2015) demonstrates the broad effectiveness of targeted and general vocalization surveys for anuran habitat use on the Crane Trust's properties. The USGS has a standard amphibian monitoring program, focused on anuran (frog and toad) vocalizations to detect the presence and absence of species as well as their relative abundance. Therein, abundance is broken up into 3 differentiable categories (Weir and Mossman 2005; USGS 2016; Table 8):

Table 8. Amphibian calling index

1	Individuals can be counted; there is space between calls
2	Calls of individuals can be distinguished but there is some overlapping of calls
3	Full chorus, calls are constant, continuous and overlapping

Surveys are to be conducted as early as 30 minutes after official sunset in humid and low wind (<15 mph) conditions, with early spring temperatures above 42° F (March-15 to May-15) and late spring-summer temperatures above 50° F (May-15 forward) (Weir and Mossman 2005; USGS 2016). Surveys should last 5 minutes per site and be conducted at least 2-4 times per survey season at each site, with 1-2 surveys conducted in the early spring period (March 15th – May 15th) and 1-2 surveys conducted in the late spring-summer period (May 16th – July 31st) at each site (USGS 2016). A recording device should be brought to each site to record novel calls and to provide evidence for species not previously detected on the Crane Trust properties. Novel calls can be investigated physically following timed survey periods at a set location. Playback can also be used to look for rare species following the official survey, but not during. A count estimate (e.g., 9 individuals) should be recorded along with the calling index (0-3); this will help us determine what the index means for each species as they vary widely in their calling habits and detection distances. If the species is too numerous to count, please record "TNTC". In 2020 we began categorizing whether vocalization activity was present locally (within 20 m of the observer) or simply present at the landscape-level. This is operationalized as a binary variable in our database with "1" indicating local activity and a "0" indicating detectable activity only at the landscape-level.

We chose 12 monitoring sites (survey sequence ordered randomly) based on a few different criteria. In 2015 we began cooperating with the Platte Basin Timelapse Project (PBT; 2018) and the Center for Global Soundscapes (CGS; 2018) to utilize passive monitoring equipment. Custom built time-lapse camera systems, as well as SM2 and SM4 wildlife recorders (Wildlife Acoustics 2018), are used to record and monitor biological activity in wetlands on Mormon and Shoemaker Islands including phenology, biophony, plant growth, water inundation, and other measures (Brinley Buckley 2016; Brinley Buckley et al. 2017; Brinley Buckley et al. 2021). In-field counts of anurans could be analyzed along with measures such as green up (NDVI/GCC), water inundation, and acoustic indices of biophony, including the Acoustic Complexity Index (ACI) (Pijanowski et al. 2011). Secondly, the Crane Trust is home to some of the most biodiverse wet meadow systems left in the Central Platte River Valley which support a diversity of anurans (Meyer et al. 2008; Ramirez and Weir 2010; Geluso and Harner 2013; Brinley Buckley et al. 2021; Malzahn et al. 2021). Woodland expansion may also provide habitat for eastern species extending their distributions west (McLean et al. 2016; Malzahn et al. 2021). Woodland and wet meadow systems at the Crane Trust host long-term monitoring transects, where vegetation, groundwater, and soils data are gathered on a periodic basis allowing for the linkage of site conditions to anuran abundance. Herpetofauna are one of the first environmental indicators of detrimental changes to wetland habitats (Price et al. 2007; Niemi et al. 2007). Monitoring will ensure that global changes are documented on a local level and will allow us to assess the various impacts of management practices over long periods of time.

Table 9. Anuran species historically detected and potentially present on Crane Trust properties

Common Name	Scientific Name	Habitat Preferences	Abundance
Plains Leopard Frog	*Lithobates blairi* (PLF)	Mesic Grasslands and Sloughs	Common
Boreal Chorus Frog	*Pseudacris maculata* (BCF)	Mesic Grasslands and Sloughs	Common
Woodhouse's Toad	*Anaxyrus woodhousii* (WT)	Widespread	Common
Bullfrog	*Lithobates catesbeianus* (BF)	Ponds and Sloughs	Common
Cope's Gray Treefrog	*Hyla chrysoscelis* (CGT)	Mesic Woodlands and Grasslands	Rare
Blanchard's Cricket Frog	*Acris crepitans blanchardi* (CR)	Sandy Areas (Platte River)	One Recording[‡]
Great Plains Toad	*Anaxyrus cognatus* (GPT)	Mesic Grasslands	Unconfirmed
Northern Leopard Frog	*Lithobates pipiens* (NLF)	Mesic Grasslands	Unconfirmed
Plains Spadefoot Toad	*Spea bombifrons* (PST)	Grasslands with Sandy Soils	Rare*

Notes: *Confirmed 2018; [‡]A. Caven recognized a Blanchard's Cricket Frog calling on one of B. Ostrom's recordings from her Master's Thesis project in 2022 (the recording is likely from 2021 or earlier); Species abundance was estimated from incidental detections and from data published by Jones et al. (1981) and Geluso and Harner (2013).

Table 10. Anuran call survey locations and associated research projects at those sites.

Run	Location	Latitude	Longitude	Associated Research
Wild Rose	Plover Pond	40.783007°	-98.473548°	PBT/CGS (<200m), Mon. Plot (RBM1)
Wild Rose	SM1	40.784215°	-98.467895°	Mon. Plot
Wild Rose	PD1	40.790665°	-98.451714°	Mon. Plot
Wild Rose	BS1	40.798032°	-98.441568°	Mon. Plot (BS1<200m), Slough Fish, H2O (East Caddis Trans.)
Wild Rose	CP Slough	40.79196°	-98.46103°	H2O (HQ Trans.), Slough Fish
Wild Rose	Ruge Pond	40.789959°	-98.492436°	Mon. Plot (R2<200m)
Mormon	PBT Slough	40.800322°	-98.417142°	PBT/CGS, Mon. Plot (NWM2<200m), H2O (Mormon Middle Trans. <200m)
Mormon	NEM2	40.800215°	-98.407738°	Mon. Plot, H2O (Mormon Middle Trans. <200m)
Mormon	RBM4	40.793912°	-98.400413°	Mon. Plot (RBM4 & SEMW3<200m)
Mormon	River Pond	40.790838°	-98.411439°	Herpetofauna Cover Board Array
Mormon	SEM1	40.790063°	-98.413953°	Mon. Plot
Mormon	Caddis	40.80779°	-98.38397°	H2O (Type Locality Slough Trans.)

Notes: PBT = Platte Basin Timelapse Project site; CGS = Center for Global Soundscapes site; Mon. Plot = Monitoring plot present (vegetation, avian, etc.); Slough Fish = Slough fish monitoring site; H2O = Transducer site (i.e., water-level logger).

Figure 15. Map of Anuran call survey locations (Top: Shoemaker, Bottom: Mormon). The purple line indicates the driving route for each survey run.

References

Allen, R.P., and J.A. Livingston. 1952. The whooping crane. National Audubon Society, New York, New York, USA, 246 pp.

Brinley-Buckley, E.M. 2016. Applications of time-lapse imagery for monitoring and illustrating ecological dynamics in a water-stressed system. Thesis. University of Nebraska-Lincoln, Lincoln, Nebraska, USA.

Brinley-Buckley, E.M., C.R. Allen, M. Forsberg, M. Farrell, and A.J. Caven. 2017. Capturing change: the duality of time-lapse imagery to acquire data and depict ecological dynamics. Ecology and Society 22(3):30.

Brinley Buckley, E.M., B.L. Gottesman, A.J. Caven, M.J. Harner, B.C. Pijanowski. 2021. Assessing ecological and environmental influences on boreal chorus frog (*Pseudacris maculata*) spring calling phenology using multimodal passive monitoring technologies. Ecological Indictors 121:107171.

Caven, A.fJ., J. Salter, and K. Geluso. 2017. *Opheodrys vernalis* (*Liochlorophis vernalis*) (Smooth Greensnake). Fire Mortality and Phenology. Herpetological Review 48(4):864-865.

Caven, A.J., K.D. Koupal, D.M. Baasch, E.M. Brinley Buckley, J. Malzahn, M.L. Forsberg, and M. Lundgren. 2021. Whooping Crane (*Grus americana*) family consumes a diversity of aquatic vertebrates during fall migration stopover at the Platte River, Nebraska. 81(4):592–607.

Center for Global Soundscapes. Accessed 2018. https://centerforglobalsoundscapes.org/

Chavez-Ramirez, F. and E. Weir. 2010. Wet meadow literature and information review. Report submitted to the Platte River Recovery Implementation Program. The Platte River Whooping Crane Habitat Maintenance Trust, Wood River, NE, USA.

Ferraro, D. Accessed 2018. A Guide to Snakes, Turtles, Frogs, Lizards, and Salamanders of Nebraska. School of Natural Resources, University of Nebraska-Lincoln, Lincoln, Nebraska, USA. http://snr.unl.edu/herpneb/

Geluso, K., B.T. Krohn, M.J. Harner, and M.J. Assenmacher. 2013. Whooping Cranes consume plains leopard frogs at migratory stopover sites in Nebraska. The Prairie Naturalist 45:91-93.

Geluso, K. and M.J. Harner. 2013. Reexamination of Herpetofauna on Mormon Island, Hall County, Nebraska, with Notes on Natural History. Transactions of the Nebraska Academy of Sciences 33:7–20.

Harner, M.J., A.J. Nelson, K. Geluso, and D.M. Simon. 2011. Chytrid fungus in American Bullfrogs (*Lithobates catesbeianus*) along the Platte River, Nebraska, USA. Herpetological Review 42:549-551.

Harner, M.J., J.N. Merlino, and G.D. Wright. 2013. Amphibian chytrid fungus in Woodhouse's toads, plains leopard frogs, and American bullfrogs along the Platte River, Nebraska, USA. Herpetological Rcvicw 44(3):459-461.

Jones, S.M., R.E. Ballinger, and J.W. Nietfeldt. 1981. Herpetofauna of Mormon Island Preserve Hall County, Nebraska. Prairie Naturalist 13(2): 33-41.

Malzahn, J.M., A.J. Caven, S. Warren, B.L. Ostrom, and D.M. Ferraro. 2021. Habitat associations and activity patterns of herpetofauna in the Central Platte River Valley, Nebraska, with notes on morphometric characteristics. Transactions of the Nebraska Academy of Sciences 41:88–105.

McLean, R.P., G.D. Wright, and K. Geluso. 2015. Cope's Gray Treefrog (*Hyla chrysoscelis*) along the Platte River, Hall County, Nebraska. Collinsorum 4(1):2-4.

Meyer, C.K., S.G. Baer, and M.R. Whiles. 2008. Ecosystem recovery across a chronosequence of restored wetlands in the Platte River Valley. Ecosystems 11(2):193-208.

Niemi, G.J., J.R. Kelly, and N.P. Danz. 2007. Environmental indicators for the coastal region of the North American Great Lakes: introduction and prospectus. Journal of Great Lakes Research 33(sp3):1-12.

Platte Basin Timelapse Project. Accessed 2018. http://plattebasintimelapse.com/

Price, S.J., R.W. Howe, J.M. Hanowski, R.R. Regal, G.J. Niemi, and C.R. Smith. 2007. Are anurans of Great Lakes coastal wetlands reliable indicators of ecological condition. Journal of Great Lakes Research 33(sp3):211-223.

Pijanowski, B.C., L.J. Villanueva-Rivera, S.L. Dumyahn, A. Farina, B.L. Krause, B.M. Napoletano, S. Gage, and N. Pieretti. 2011. Soundscape ecology: the science of sound in the landscape. BioScience 61(3):203-216.

Schultz, C., and A.J. Caven. 2021. Lined Snake (*Tropidoclonion lineatum*) Prescribed Fire Mortality. Transactions of the Nebraska Academy of Sciences 41:42–45.

[USGS] US Geological Survey. 2016. North American Amphibian Monitoring Program. Eastern Ecological Science Center, Kearneysville, West Virginia, USA. <https://www.usgs.gov/centers/eesc/science/north-american-amphibian-monitoring-program>

Weir, L.A. and M.J. Mossman. 2005. North American amphibian monitoring program (NAAMP). Pages 307-313 *in* M. Lannoo, editor, Amphibian Declines: The Conservation Status of United States Species. University of California Press, Berkeley, California, USA, 1115 pp.

Wiese, J. D., and A. J. Caven. 2017. *Tropidoclonion lineatum* (Lined Snake), *Thamnophis sirtalis* (Common Gartersnake). Refugia and Mortality. Herpetological Review 48(4): 868-869.

Wiese, J.D., A.J. Caven, and E.M. Brinley-Buckley. 2016b. Eastern Racer (*Coluber constrictor*) mortality as a result of early emergence from a man-made structure hibernaculum in South-central Nebraska. Collinsorum 5(1):3-5.

Wiese, J.D., K.C. King, and A.J. Caven. 2016a. The utilization of senesced wetland plant material by *Thamnophis sirtalis* as a thermoregulation microsite in a flooded system. Collinsorum 5(4):12-14.

Wiese, J.D., K.C. King, A.J. Caven, and N. Arcilla. 2017b. Winter predation of an adult Spiny Softshell (*Apalone spinifera*) likely committed by a Bald Eagle (*Haliaeetus leucocephalus*) in central Nebraska. Collinsorum 6(1):14-19.

Wiese, J.D., B. Krohn, and A.J. Caven. 2017a. Common Gartersnake (*Thamnophis sirtalis*) Mortality likely resulting from cold exposure following late winter hibernaculum emergence. Collinsorum 6:15-16.

Wiese, J.D., E.D. Plock, and K. Geluso. 2016c. Common Gartersnake (*Thamnophis sirtalis*) mortality due to haying practices in South-central Nebraska. Collinsorum 5(4):15-16.

Wildlife Acoustics (Accessed 2018). Song Meter SM4. <https://www.wildlifeacoustics.com/products/song-meter-sm4>

Anuran Call Study Survey Datasheet

Date: _____   Observer(s): _____

Location	Start/End Time	Species	Call Index	Count Est.	Local (1,0)	Temp.	Humid.	Wind Spd.	Notes

Chapter 9: Greater Prairie Chicken Monitoring Protocol

Project Goals

The Greater Prairie Chicken (*Tympanuchus cupido pinnatus*) has declined throughout much of its range within the last century (Svedarsky et al. 2000). In fact, the most secure remaining populations actually persist west of their historic range in what is an adopted range within the central Great Plains (Svedarsky et al. 2000). This species requires large relatively contiguous expanses of grassland in various stages of succession (e.g., intensively grazed, moderately grazed, and rested) to complete their annual life cycle (e.g., lekking, brood rearing, and nesting) and therefore have been considered an indicator of grassland condition and ecological function (Winter and Faaborg 1999, Robb and Schroeder 2005). Greater Prairie Chickens can be residents or short-distance migrants regionally (Johnson et al. 2020). Though wintering Greater Prairie Chickens have been consistently detected in the Central Platte River of Nebraska since the early 1980s, evidence of breeding behavior has not been recorded until the last decade (Lingle and Hay 1982). Lekking behavior was originally recorded on Mormon Island during a warmer than average January in 2015 and breeding activity continued through the spring months, however, the ratio of males to females was highly imbalanced indicating a newly establishing population (Caven et al. 2018). Since 2015 we have monitored the main lek in some capacity in most years through scan sampling, acoustic monitoring, and/or camera traps (Caven et al. 2018, King et al. 2018). During the course of our study additional smaller and more ephemeral leks have also become established on Mormon Island, but the main lekking site near the central cattle watering tank has received consistent use annually (Caven et al. 2018). We updated our monitoring protocols in 2019 and formalized the processing for our camera trap data and scan sampling procedures to improve the sensitivity of monitoring efforts. Instantaneous scan sampling procedures followed Altmann (1974), and behavioral interpretation was based on Hamerstrom (1939), Hamerstrom and Hamerstrom (1973), and Johnsgard (2016). We also increased the consistency with which we recorded acoustic monitoring data.

We continue to have multiple objectives for monitoring this lek. First, we would like to continue to document the establishment of a new breeding population of Greater Prairie Chickens in the CPRV and on Mormon Island. For this reason, we will be closely tracking the number of females attending the lek as an indicator of population establishment and possibly of success. Secondly, given the unique phenology of the lek, which generally begins full displays before many leks to its south (Caven et al. 2018), we would like to continue to track its chronology in relation to environmental conditions. Has this lek simply established an aberrant cultural pattern, is this behavior ephemeral and simply related to warmer than average winter weather, or is this behavior ultimately driven by climatic shifts? To answer such a question long-term data will be important. A related objective is to track male behavioral patterns in relation to weather variables controlling for female presence on the lek site, which is known to increase activity levels (Raynor et al. 2017). Another objective for monitoring this lek is to document the Great Prairie Chicken's behavior in relation to conspecifics in an understudied portion of their range that they are actively recolonizing. A primary goal is to document their interactions with and responses to predators as well as closely related conspecifics and other grassland birds. To date we have documented novel depredation attempts from Prairie Falcons (*Falco mexicanus*) and a rare occurrence of a Sharp-tailed Grouse (*Tympanuchus phasianellus*) lekking with the Greater Prairie Chickens on the southern edge of their range (Caven et al. 2017, King et al. 2018).

Finally, in-person monitoring is time-intensive and can be influenced by observational bias. Therefore, we have integrated a number of additional monitoring tools to improve the temporal resolution of our data. We can also use these tools, including acoustic data as well as time-lapse and motion sensor camera trap data, to supplement, evaluate, and potentially validate our in-person survey data and vice versa (Raynor et al. 2017).

Project Methods

Our current plan is to implement the updated scan sampling approach outlined in the datasheet (Appendix 10) weekly and to continue to regularly record acoustic data as well as camera trap data at the main perennial lek site on Mormon Island throughout the active lekking season from mid-January (~15th) through late-May (~21st) annually. Recommended settings include audio recorders set at an interval of 5 minutes at the top of every hour from sunrise to late morning (~3 hrs.) daily and camera traps set to motion sensor and time lapse, with pictures being captured every 15 minutes for 3 hours after sunrise and before sunset. We should reevaluate our monitoring approach periodically after data examination (every ~3 years), but the condition of this lek should be tracked indefinitely as it represents a unique resource for Mormon Island and the CPRV at large. It may well represent the founding lek of which may ultimately become a CPRV-wide breeding metapopulation as recent reports (<3 years) indicate additional lekking populations on Shoemaker Island and surrounding conservation landscapes.

Data will be recorded in the field on the Greater Prairie-Chicken lekking observations datasheet. These sheets should be scanned and saved in the GRPC Scanned Datasheets folder on the Company X drive, located within the GRPC Lekking Observations and Data 2019-Onward folder under the Greater Prairie Chicken folder, the Avian Monitoring folder, and the Science Program folder. All information on the datasheets will also be entered into the GRPC Lekking Observation Database (2019-Onward). Photos from the camera trap located on the main perennial lek site on Mormon Island will be located within the GRPC Camera Trap Data 2019-Onward folder on the Company X drive under the Greater Prairie Chicken folder. Three main folders exist to help with organization, including a Quality Photos folder (which houses photos of great quality or novel occurrences), a Predator Photos folder (which houses photos with predators present on the lek), and an Unprocessed Photos folder. The Unprocessed Photos folder contains unchecked photos that need to be added to the GRPC Camera Trap Database. Within this database, all identifiable GRPC from the pictures are counted once, as doubles may occur due to panoramic combination. If the camera produces multiple photos for one time slot (i.e., 6:50 AM), then the highest count is to be recorded in the camera trap related database.

References

Altmann, J. 1974. Observational study of behavior: sampling methods. Behaviour 49:227-267.

Caven, A.J., J.D. Wiese, and W.R. Wallauer. 2017. Prairie Falcon depredation attempts on a Greater Prairie-chicken lek in south-central Nebraska. The Prairie Naturalist 49(2):76-78

Caven, A.J., K.C. King, E.M. Brinley Buckley, G.D. Wright, N. Arcilla, and R.P. McLean. 2018. Sustained early Interior Greater Prairie-chicken (*Tympanuchus cupido pinnatus*) lekking behavior at lek in central Nebraska. Kansas Ornithological Society Bulletin 69(3):29-40.

Hamerstrom, F.N., Jr. 1939. A study of Wisconsin greater prairie-chicken and sharp-tailed grouse. Wilson Bulletin 51:105-120.

Hamerstrom, F.N., Jr. and F. Hamerstrom. 1973. The greater prairie-chicken in Wisconsin – highlights of a 22-year study of counts, behavior, movements, turnover, and habitat. Wisconsin Department of Natural Resources. Technical Bulletin Number 64.

Johnsgard, P.A. 2016. The North American Grouse: Their Biology and Behavior. Zea E-Books, University of Nebraska-Lincoln Libraries, Book 41.

Johnson, J.A., M.A. Schroeder, and L.A. Robb. 2020. Greater Prairie-Chicken (Tympanuchus cupido), version 1.0. In A.F. Poole, editor, Birds of the World. Cornell Lab of Ornithology, Ithaca, NY, USA. <https://doi.org/10.2173/bow.grpchi.01>

King, K.C., A.J. Caven, and K. Geluso. 2018. Lekking behavior of a Sharp-tailed Grouse in south-central Nebraska. The Prairie Naturalist 50(1):39-41.

Lingle, G.R., and M.A. Hay. 1982. A checklist of the birds of Mormon Island Crane Meadows. Nebraska Bird Review 50:27-36.

Raynor, E.J., C.E. Whalen, M. Bomberger Brown, and L.A. Powell. 2017. Location matters: evaluating Greater Prairie-Chicken (*Tympanuchus cupido*) boom chorus propagation. Avian Conservation and Ecology 12(2):17. <https://doi.org/10.5751/ACE-01126-120217>

Robb, L.A., and M.A. Schroeder. 2005. Greater Prairie-Chicken (*Tympanuchus cupido*): a technical conservation assessment. USDA Forest Service, Rocky Mountain Region, Lakewood, CO, USA, 79 pp.

Svedarsky, W.D., R.L. Westemeier, R.J. Robel, S. Gough, and J.E. Toepfer. 2000. Status and management of the Greater Prairie-Chicken *Tympanuchus cupido pinnatus* in North America. Wildlife Biology 6:277–284.

Winter, M. and J. Faaborg. 1999. Patterns of area sensitivity in grassland-nesting birds. Conservation Biology 13:1424-1436.

Greater Prairie-Chicken Lekking Observations – Scan Sampling Data Sheet 2022

Survey Information   Observer: _____ Recorder: _____
Date: _____ Start Time: _____ End Time: _____ Wind (mph): _____ Temp.: _____ % Clouds: _____
Max Count GRPC: _____ Males: _____ Females: _____ Latitude (if changed): _____ Longitude: _____

Lek Site Description
Veg. height (cm): _____ Plant Regrowth ("green-up"; Y/N): _____ Management (hay, burn, graze, mow): _____
Management Description (acres, etc.):
Habitat Conditions (flooding, vegetation spp., etc.):

Interspecific Interactions
Species: _____ Interaction Start: _____ Int. End: _____ Physical Contact (Y/N): _____ Success (Y/N/U): _____
GRPC Reaction (count), Flush: _____ Crouch: _____ Fight: _____ Walk Away: _____ None: _____
Hunting Behavior (tail chase, dive, run through, etc.):
Event Description (PRFA pursued GRPC for ~1 km, etc.):

Scan Time	♂	♀	Display					Copulation	Foraging	Loafing	Flying	Predation Avoidance
			Booming	Strutting	Flutter Jump	Nuptial Bow	Sparring					

Notes: Complete a scan every 3-mintues and enter the total number of birds doing each behavior at the time of the scan into the appropriate box (instantaneous scan sampling). Observe Greater Prairie Chickens for a minimum of 30 minutes and a maximum of 1 hr. **Booming:** Deep three-syllable call produced as esophageal air sacs inflate and deflate. **Strutting:** Males bend forward positioning bodies parallel to the ground with wings to the sides and perform a rapid quickstep. **Flutter Jumps:** Leap into air while flapping wings, often producing whoop, cackle, or whine calls, to display position in lek. **Nuptial Bow:** Male spreads wings and lowers bill to the ground with pinnae in erect position; often precedes copulation. **Sparring:** Dominance displays through charging and fighting. **Copulation:** Male mounts female. **Foraging:** Includes movement foraging (walking, scanning, and occasionally peeking at the ground). **Loafing:** Includes preening and resting. **Flying:** Record all flights regardless of distance. **Predation Avoidance:** Flushing, crouching, fighting back, or walking away from potential predators. **An individual can be completing more than 1 behavior at a time, tally their behaviors in all applicable categories. Add a count for females but do not tally behavior.**

Chapter 10: Whooping Crane Diurnal Behavior and Natural History during Migration

Project Goals

The objective of this study is to collect behavioral data that allows us to calculate Whooping Crane time budgets and link them to the habitats they are utilizing (Lingle et al. 1991; Jorgensen and Dinan 2016). We use an "instantaneous scan sampling" approach which includes counting the number of Whooping Cranes displaying a particular behavior at one-minute intervals for a period of no less than 30 minutes (Altmann 1974). This behavioral monitoring can help us determine which values various habitats provide (i.e. – forage resources, safe areas for social display, etc.) as well as how behavior varies within and across habitat types. This data can also help us document potential threats (e.g. – frequency of attempted depredation by Bald Eagles; Rabbe et al. 2019) as well as specific forage resources (e.g. – Channel Catfish; Caven et al. 2019a). In short, we will gather natural history information that has the potential to inform conservation efforts through behavioral observations. For instance, recent research has found that Whooping Cranes consume a wider variety of food items during migration than previously documented, including a variety of wetland-dependent vertebrates (Geluso et al. 2013; Caven et al. 2019a). Behavioral surveys will be paired with and serve as a supplement to regular efforts to confirm public sightings of Whooping Cranes in the Platte River Valley and beyond for the federally managed "U.S. Fish and Wildlife Service (USFWS) public sightings database" (Lewis 1992; Caven et al. 2020). Additional support for science-focused staff in the months of March and April (spring migration) as well as October and November (fall migration) will help us scale-up the collection of behavioral data as well as increase ground crew efforts to locate/relocate Whooping Cranes, thereby further improving the USFWS public sightings database. Scaled up efforts will include having a full-time staff, interns, and/or contractors track diurnal Whooping Crane behavior during the migration seasons.

Project Methods

Locations will be provided via the USFWS-managed public sightings database, to which the Crane Trust often contributes locally. Qualified biologists will be sent into the field to confirm public reports of Whooping Cranes and, in addition to filling out a traditional sightings report, biologists will also conduct scan sampling to get a more systematic view of behavior. Additionally, biologists will be sent to the locations of some Whooping Cranes tracked with cellular technology to document behavior with the goal of linking this behavior to movements measured via new-age tracking technologies. Research will be conducted predominantly in south-central Nebraska (Rainwater Basins, the Loup River system, Platte River system, etc.) with occasional work outside of this area (throughout Nebraska, northern Kansas, etc.) as time and funds allow. Work will be conducted following the guidelines drafted by the USFWS and the Nebraska Game and Parks Commission (NGPC) for "avoiding Whooping Crane disturbance and harassment" including making observations from >610 m (~0.4 mi, 2,000 ft.), avoiding intrusions into habitats until after the cranes have clearly departed the area to measure habitat parameters etc., and immediately reporting any information regarding observations of injured cranes to the proper authorities. The only case in which research staff would be closer than 610 m to a Whooping Crane would be if an individual or group approached an observing biologist concealed in a blind or vehicle. In this case the biologist will remain in the blind until 30 minutes after dusk or the cranes have departed or moved far enough away to allow the biologist to depart without disturbing the Whooping Cranes.

At each site, time, date, and weather conditions (wind, cloud cover, temp, etc.) will be recorded along with basic locational (description, latitude, longitude), habitat, and land management data. We have worked to create a discrete list of habitat types (e.g., lowland tallgrass prairie, shallow marsh, cornfield, etc.) that is detailed in the attached datasheet and full-page figure providing visual and narrative descriptions of prairie and wetland habitats. We also included a section to record pertinent notes on habitat characteristics. We created a list of categories that apply to management in herbaceous and agricultural systems (e.g., grazed, burned, hayed, harvested, etc.), and provide a space on the datasheet for detailed notes regarding management as well. We will measure the distance from Whooping Crane locations to water (0 = within standing water) as well as major rivers (only in river valleys) using a range finder in the field for shorter measurements, and the most recent aerial imagery available from the same season and climatic conditions for longer measurements (e.g., wet spring, etc.). We will also measure the unobstructed wetted width of wetland habitats used by Whooping Cranes. Unobstructed wetted width (UOWW) will include the total width of the palustrine/lacustrine wetland or river channel unobstructed by vegetation >1.5 m in height (Pearse et al. 2017; Caven et al. 2019b). Wetlands will be measured across their narrowest full width whereas rivers will be measured perpendicular to their banks. Water depth will be estimated based on the degree to which the tarsus is submerged in water (mean tarsus length = 28 cm; Johnsgard 1983; Caven et al. 2019a). Finally, we will record the bridge segment of Whooping Crane locations for those sites in the Central Platte River Valley (CPRV; 1-11; Caven 2019b).

We will also record the physical description of observed Whooping Cranes, including bands, other distinctive physical characteristics, and any injuries. In addition to rows for recording 40 minutes of systematic scan sampling data, our datasheet provides space to narratively describe interesting observations or contextualize behavioral data. For instance, scan sampling data can indicate that 90% of time was spent exhibiting alert-defensive behavior, but the narrative portion will allow biologists to describe the contexts under which such behavior arises. For example, maybe a coyote approached via a nearby wetland bank, etc. We include a separate datasheet with which to further document eagle-crane interactions considering the recent increase in observations of Bald Eagles attempting to depredate crane species regionally (Rabbe et al. 2019). The crane-eagle interactions datasheet represents a stand-alone protocol that also applies to Sandhill Cranes and thus will have some overlap in questions (e.g., distance to woodland) with Whooping Crane behavioral scan sampling. We will also record the presence of any aircraft, its altitude estimated visually (max = 1,500 m), the type of aircraft, and Whooping Crane reactions. Again, we provided a discrete list of potential reactions as well as space to provide a narrative description. Finally, we will note each use location's distance to the nearest powerline and the powerline type (major >5 lines, minor <5) as well as distance to the nearest paved road.

We will rely on high resolution long-range photography and videography to documented Whooping Crane foraging behavior using a Tamron SP 150-600 mm lens paired with a Nikon DSLR Camera as well as a Nikon Coolpix P1000 Super-telephoto digital camera (3,000 mm zoom equivalent). To ensure we do not disturb Whooping Cranes, flash photography will never be used, and photographs will only be taken under natural light conditions. Our goal will be to collect a minimum of 30 minutes of scan sampling data, given the Whooping Cranes continued presence. However, if at any time during that 30 minutes biologists observe a Whooping Crane consuming visually discernable food items through the spotting scope, scan sampling will be paused to focus on shooting photographs of the diet items considering the sparse amount of information available concerning diet regionally and during migration generally (Caven 2019a). Following photography of visually discernable diet items, biologists will resume behavioral scan sampling until at least 30 minutes of total effort is reached. Following the completion of 30 minutes of scan sampling, the observing biologist will assess whether to continue based on several criteria including the number of other birds to observe locally, the novelty of behaviors being recorded, and the degree to which the observer is safely and comfortably concealed from its subjects to ensure no disturbances to migrating Whooping Cranes. In addition to documenting diet items, camera equipment will be used for long-range videography, specifically to collect 10 minutes of video following tracked birds to facilitate the evaluation of their on-the-ground behavior in comparison to accelerometer data from tracking devices. It will be extremely important to record the start and end times of the video to the nearest second to allow for direct comparison between photographic and accelerometer data. Furthermore, it will be critical to maintain focus on a single banded and tracked crane while shooting video.

Sandhill Crane Breeding Behavioral Observations

The Crane Trust developed a very similar behavioral research protocol to the Whooping Crane diurnal behavior study to track late spring and early summer Sandhill Cranes activity in the Central Platte River Valley in 2018 (Malzahn et al. 2018). Given the comparable behavioral repertoires of Sandhill Cranes and Whooping Cranes, the behavior categories from both studies are relatively similar despite differing project objectives (Ellis et al. 1998; Appendix 15). Therefore, the Sandhill Crane breeding season observations study, which will likely be more intermittent than long-term, is presented here as a heading under chapter 10. Malzahn et al. (2018) provides a detailed description of the protocol used to assess potential Sandhill Crane breeding season activity and the datasheet is presented as Appendix 15 if Sandhill Cranes are believed to be breeding regionally in the near future.

References

Altmann, J. 1974. Observational study of behavior: sampling methods. Behaviour 49:227-267.

Caven, A.J., J. Malzahn, K.D. Koupal, E.M. Brinley Buckley, J.D. Wiese, R. Rasmussen, C. Steenson. 2019a. Adult Whooping Crane (*Grus americana*) Consumption of Juvenile Channel Catfish (*Ictalurus punctatus*) during the Avian Spring Migration in the Central Platte River Valley, Nebraska, USA. Monographs of the Western North American Naturalist 11:14-23.

Caven, A.J., E.M. Brinley Buckley, K.C. King, J.D. Wiese, D.M. Baasch, G.D. Wright, M.J. Harner, A.T. Pearse, M. Rabbe, D.M. Varner, B. Krohn, N. Arcilla, K.D. Schroeder, K.F. Dinan. 2019b. Temporospatial shifts in Sandhill Crane staging in the Central Platte River Valley in response to climatic variation and habitat change. Monographs of the Western North American Naturalist 11(1):33-76.

Caven, A.J., M. Rabbe, J. Malzahn, and A.E. Lacy. 2020. Trends in the occurrence of large Whooping Crane groups during migration in the great plains, USA. Heliyon 6(4):e03549.

Ellis, D.H., S.R. Swengel, G.W. Archibald, and C.B. Kepler. 1998. A sociogram for the cranes of the world. Behavioural Processes 43:125–151.

Geluso, K., B.T. Krohn, M.J. Harner, and M.J. Assenmacher. 2013. Whooping Cranes consume Plains Leopard Frogs at migratory stopover sites in Nebraska. The Prairie Naturalist 45:91-93.

Johnsgard, P.A. 1983. Cranes of the World. Indiana University Press, Bloomington, Indiana, USA, 279 pp.

Jorgensen, J.G., and L.R. Dinan. 2016. Whooping Crane (*Grus americana*) behavior, habitat use and wildlife watching visitation during migratory stopover at two wildlife management areas in Nebraska 2015–2016. Nongame Bird Program, Nebraska Game and Parks Commission, Lincoln, Nebraska, USA, 26 pp.

Lewis, J.C. 1992. The contingency plan for federal-state cooperative protection of whooping cranes. Pages 295-300 *in* D.A. Wood, editor, Proceedings of the 1988 North American Crane Workshop, 22–24 February 1988, Lake Wales, Florida. State of Florida Game and Fresh Water Fish Commission, Tallahassee, Florida, USA. Nongame Wildlife Program Technical Report No. 12.

Lingle, G.R., G.A. Wingfield, and J.W. Ziewitz. 1991. The migration ecology of Whooping Cranes in Nebraska, USA. Pages 395-401 *in* J.T. Harris, editor, Proceedings of the 1987 International Crane Workshop, 1-10 May 1987, Qiqihar, China. International Crane Foundation, Baraboo, Wisconsin, USA.

Malzahn, J., A.J. Caven, M. Dettweiler, and J.D. Wiese. 2018. Sandhill Crane Activity in the Central Platte River Valley in Late May and Early June. The Nebraska Bird Review 86(4):175-180.

Pearse, A.T., M.J. Harner, D.M. Baasch, G.D. Wright, A.J. Caven, and K.L. Metzger. 2017. Evaluation of nocturnal roost and diurnal sites used by Whooping Cranes in the Great Plains, USA. U.S. Geological Survey Open-File Report 2016–1209, 29 pp.

Rabbe, M.R., A.J. Caven, and J.D. Wiese. 2019. First description of a Bald Eagle (*Haliaeetus leucocephalus*) attempting depredation on an adult Whooping Crane (*Grus americana*) of the Aransas-Wood Buffalo population. Monographs of the Western North American Naturalist 11(1):24-32.

Whooping Crane Migration Behavior Observations—Scan Sampling Data Sheet
*To be paired with USFWS Whooping Crane Report Field Sheet. Please send scanned datasheet to Andrew J. Caven, acaven@cranetrust.org, 6611 W Whooping Crane Dr., Wood River, NE 68883.

Observer:_____ Date:_____ St. Time: _____ End Time: _____ Wind (mph): _____

Temp (°F): _____ % Clouds: _____ No. Adult: _____ No. Juv.:_____ Dist. H2O (m): _____

Latitude: _____ Longitude: _____ Brdg. Seg. (CPRV, 1-11): _____

Dist. River (m; river valleys, e.g., Loup, etc.): _____ H2O Depth (cm; via tarsus): _____

Unobstructed Wetted Width (m; wetlands, e.g., palustrine, lacustrine, riverine): _____

Nearest Standing Water Feature (see habitat desc.): _____ Dist. Woodland (m): _____

Location Description (e.g.- 1.5km W Alda Rd, 2 km N Platte River Rd):

Mgmt. Use Loc. (grazed, burned, hayed (<1 year), rested, harvested, disked, standing crop, other): _____

Mgmt. Notes:

Habitat Description (e.g.- lowland tallgrass prairie, wet meadow, shallow marsh, deep marsh, corn, fallow, alfalfa, soybean, wheat/barley, or other agricultural field (use "flooded" as modifier if needed), river, stream, open-water slough, drainage ditch/canal, open-water palustrine, natural lacustrine, pond/pit, reservoir, other):_____

Habitat Notes:

Physical Description/Evidence of Injury (e.g.- YL/YL right leg, left wing not extending fully, etc.):

Overall Behavior Description (e.g.- 2 adult WHCRs observed foraging and preening, etc.):

Aircraft Detected (Y/N): _____ Estimated Aircraft Altitude (m; max ≥1,500 m): _____

Type (Commercial Airline, Small Plane, Helicopter, etc.): _____

Whooping Crane(s) Reaction (Alert, Defensive, Flush, Nothing, etc.): _____

WHCR Reaction Notes:

Diet Items Documented (to lowest taxa, e.g., Anura, Mollusca, Arthropoda, Actinopterygii (bony fishes), Pseudacris (chorus frogs), etc.):_____ Pic. No(s).: _____

No. Consumed (of each spp., e.g., Anura – 2, etc.): _____

Foraging Notes:

Video (Y/N): _____ Bands, R:__/__ L: __/__ File No.: _____ Start (hh:mm:ss): _____ End: _____

Dist. Paved Rd. (m): _____ Dist. Pwr. Ln (m): _____ Pwr. Ln. Type (min./maj.): _____

Scan Time	Foraging	Social (Conspecific)	Social (Interspecific)	Alert-Defensive	Flying-Walking	Loafing	Preening
1.							
2.							
3.							
4.							
5.							
6.							
7.							
8.							
9.							
10.							
11.							
12.							
13.							
14.							
15.							
16.							
17.							
18.							
19.							
20.							
21.							
22.							
23.							
24.							
25.							
26.							
27.							
28.							
29.							
30.							
31.							
32.							
33.							
34.							
35.							
36.							
37.							
38.							
39.							
40.							

Notes: Count the number of Whooping Cranes displaying each behavioral category that applies during each instantaneous scan sample (what is each crane doing in a particular instant). An individual could potentially be displaying more than one behavior at a time. Try to complete a scan every 1-minute of sampling. Observe Whooping Cranes for at least 30-minutes. **Foraging**- Includes movement foraging (walking scanning and occasionally pecking at ground), time spent continuously foraging, and drinking. **Social (Conspecific)**- All dancing, pair bonding, vocalization, or aggressive interactions with other Whooping Cranes. **Social (Interspecific)**- All behavioral interactions with other species that do not pose a predatory risk (i.e. ducks, geese, etc.); often includes agonistic behavior with Sandhill Cranes or other flocking waterbirds. **Alert-Defensive**- Alarm calls, wing-spread, jump-rake, or bill-stab displays directed at potential threats (people, coyotes, eagles, etc.). **Flying/Walking** - Record all flights regardless of distance as well as walking not associated with foraging (getting from A to B directly). **Loafing**- Resting, generally while standing. **Preening**- Cleaning/oiling feathers, bathing, etc.

95

Appendix 13: Habitat classification

Habitat Classifications – Whooping Crane Behavioral Monitoring

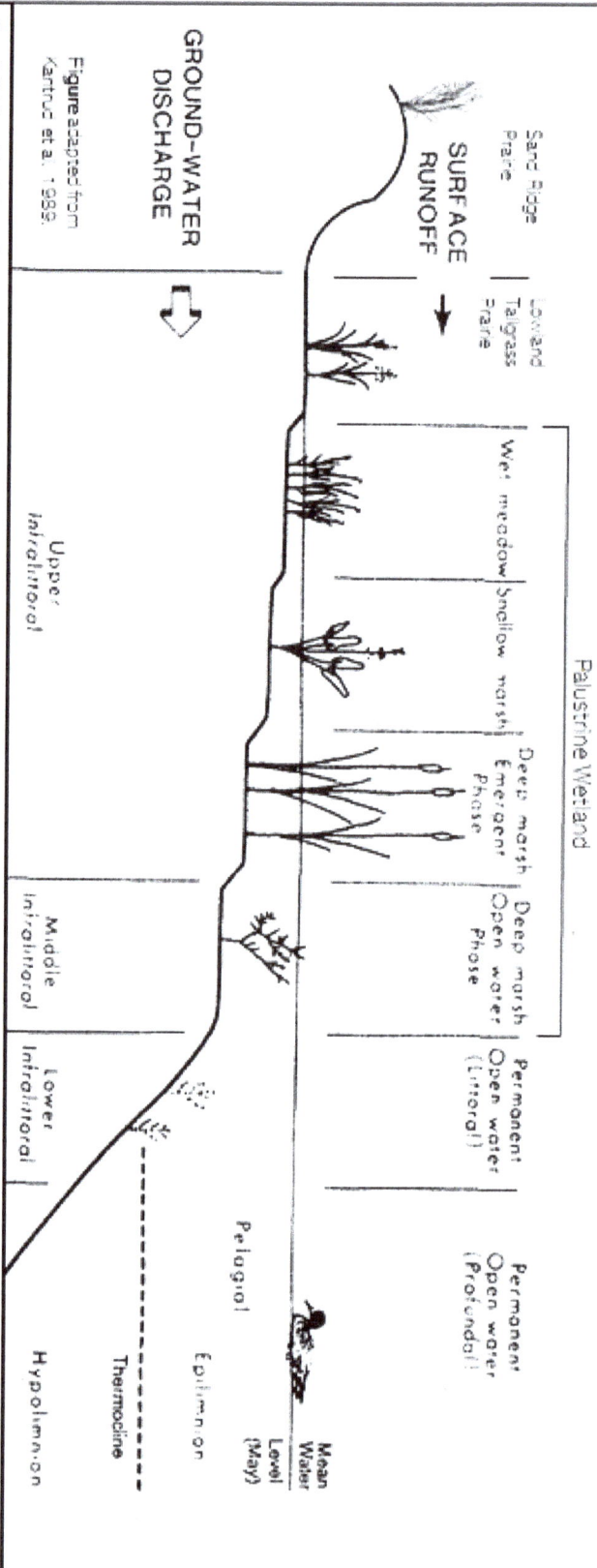

Figure adapted from Kantrud et al. 1989.

Sand Ridge Prairie exists on the highest ridges in the PRV and has more mixed grass components; it is the only herbaceous habitat locally that supports a plant community disconnected from groundwater. Dominant species include little bluestem, rough dropseed and needle-and-thread. Lowland Tallgrass Prairie represents a western extension of tallgrass prairie resultant from access to shallow groundwater. This habitat type is not a wetland and infrequently supports standing water but can flood. Dominant species include big bluestem, Indiangrass, and switchgrass, which have root systems that exceed depths of 2 m. Wet Meadows are temporarily to seasonally flooded in the spring and are dominated by sedges (Carex spp.) including Emory's sedge, wooly sedge, and other mesic graminoids including common threesquare and prairie cordgrass. Shallow Marshes are seasonally to semi-permanently flooded in the spring to early summer and regionally dominated by cattails in the PRV, with large-fruit bur-reed, and spikerushes (Eleocharis spp) also common. Deep Marshes are semi-permanent wetlands that tend to only dry up only during droughts, and even then, groundwater remains close to the surface. Dominant species include softstem bulrush, smart weeds (Polygonum spp.), and arrowhead species (Sagittaria spp.). River - flowing water occupying a relatively patterned course on a significant scale. Stream - flowing water on a more localized scale (includes creeks, etc.). Open-water Slough is a topographic distinction representing a naturally occurring linear swale that can support a range of wetland habitats (e.g. shallow marsh or open water). They are groundwater fed and generally exhibit little flow. Drainage Ditch/Canal an artificially constructed linear swale that intentionally transports water based on human needs. Open-water Palustrine wetlands are naturally occurring, more often temporary or semi-permanent, less than 8 ha in area, and less than 2.5 m in depth. Natural Lacustrine wetlands are naturally occurring permanent bodies of water that include deeper habitats (> 2.5 m) and are generally larger than 8 ha (e.g. lakes). These systems can be bordered by palustrine wetlands, but often have little vegetation on their banks as a result of wave action. Pond/Pit includes artificially created lacustrine habitats aside from reservoirs such as impoundments, excavated ponds, and gravel pits, etc. Reservoirs are artificially created lacustrine habitats (generally dammed rivers) used for holding irrigation water and recreation.

Eagle-Crane Interaction Study

Coordinated by the Crane Trust

Send Form To: acaven@cranetrust.org or 6611 W Whooping Crane Drive, Wood River, NE, 68883

CONTACT INFO

Observer(s): _____ Phone Number: _____ Email: _____

LOCATION INFO

State: _____ Latitude (decimal degrees): _____ Longitude: _____ County: _____

Nearest Town (i.e., 4.5 mi. NW Prosser, NE): _____

Landmark Description (i.e., 200 m W Alda Bridge on Platte River): _____

OBSERVATION INFO

Observation Date: _____ Obs. Time Start (CST): _____ Obs. Time End: _____

Observer distance (i.e., 400 m): _____ Photo/Video Attached (Y/N): _____

No. Eagles Present (w/in 800 m of Cranes): _____ No. Interacting with Cranes: _____

No. Cranes in Group (Circle 1):   1   2-5   6-20   21-100   101-500   501-1,000   >1,000

Interacting Eagle Ages (Count): 1st yr.: ____ 2nd yr.: ____ 3rd yr. (Bald Eagles only): ____ Adult: _____ Unknown: ____

Interacting Crane Ages (Count): Juvenile: _____ Adult: ____ Unknown: _____ Crane-Eagle Contact (Y/N/Unk.): _____

Habitat (Circle All that Apply): River   Herbaceous Wetland   Lake/Reservoir   Pasture/Meadow   Agricultural Field
Woodland/Forest   Other: _____

Habitat Description (e.g., submerged sandbar in Platte River): _____

If applicable, Estimated Water Depth: _____ Distance to Woodland (m): _____

Wetted Width (shortest distance across the center of wetland feature (m)): _____

<u>CIRCLE ONE</u>

Sky Condition:	Cloudy	Partly Cloudy	Clear	
Observation By:	Binoculars	Spotting Scope	Camera	Visual
Crane Species:	Sandhill Crane	Whooping Crane	Both	
Eagle Species:	Bald Eagle	Golden Eagle	Both	
Crane Behavior Prior:	Loafing	Feeding	Flying	Other: _____
Crane Reaction:	Defensive	Flush	None	Other: _____
Predation Success:	Yes No	Unknown, Notes: _____		
Crane Sick/Injured	Yes No	Unknown, Notes: _____		

Description of Events:

97

Sandhill Crane Breeding Season Observations—Scan Sampling Data Sheet

Date: _____ __ Start Time: _____ _End Time: _____Wind (mph):_____ Temp: _____

Percent Clouds: _____ Latitude: _____ _____ Longitude: _____

Location Description (e.g.- 1.5km W Alda Rd, 2km N Platte River Rd):

Habitat Description (e.g.- corn field, lowland prairie, reuse pit, river, wetland, etc.):

Overall Behavior Description (e.g.- 2 adult SACRs observed foraging, preening, and tending nest, etc.)

Scan Time	Foraging	Social-Mating	Preening	Loafing	Alert-Defensive	Flying	Parental

Notes: Check all categories that apply on each scan. Try to complete a scan every 3-minutes of sampling. Try to observe Sandhill Cranes for at minimum 15-minutes. **Foraging**- Includes movement foraging (walking scanning and occasionally peck at ground). **Social**- All dancing, pair bonding, or aggressive interactions with other cranes. Please pay special attention to any copulation activity. **Preening**- Cleaning/oiling feathers, bathing, etc. **Alert-Defensive**- Alarm calls, wing-spread, or bill-stab displays directed at potential threats (people, raccoons, eagles, etc.). **Flying**- record all flights regardless of distance. **Parental**- Brooding behavior, feeding chicks, etc. If feeding chicks simply count both parental and foraging.

Chapter 11: Sandhill Crane Migration Aerial Survey

Project History and Goals

The Crane Trust first began conducting weekly early morning aerial surveys of Sandhill Crane roosts in the Central Platte River Valley between Chapman and Overton, Nebraska, in 1998 (Davis 2001, 2003). Early survey efforts were focused from late February or early March to early April and used videography to assess Sandhill Crane habitat use and roost locations (Davis 2001, 2003). From 2002 to 2010 and 2013 to 2022, survey efforts did not use videography, but instead employed Global Positioning Systems (GPS) to more accurately record Sandhill Crane roosting locations (Buckley 2011; Baasch et al. 2019; Caven et al. 2019). Survey efforts generally spanned from mid-February to mid-April from 2002 to 2022. From 2016 to 2022 we incorporated a bias correction procedure that improved the accuracy of and specified a confidence interval for Sandhill Crane abundance indices (Ferguson et al. 1979; Bowman et al. 2014; Caven et al. 2019, 2020). As public interest in the Sandhill Crane migration continued to grow, we began to provide periodic and then weekly updates regarding this aerial survey project during the spring (https://cranetrust.org/news-events/the-prairie-pulse.html).

The primary objectives of these surveys are to determine the distribution of Sandhill Crane roosts, provide a reliable index of Sandhill Crane abundance per survey week, and to track changes in the chronology and distribution of Sandhill Cranes within the Central Platte River Valley across survey years (Caven et al. 2019, 2020). This survey effort also allows us to investigate Sandhill Crane habitat use in response to land management (tree removal and river disking) conducted in the river valley by the Crane Trust and other conservation partners (Davis 2003; Buckley 2011; Baasch et al. 2019; Caven et al. 2019). Additionally, the research program helps us track general trends in peak abundance over time.

Project Methods

Survey Timing

Aerial Sandhill Crane roost surveys are conducted each week from the middle of February to the middle of April for a period of 10 weeks (Table 11). However, we have tended to complete a total of 6-10 surveys each year depending on funding, Sandhill Crane presence, and long-term weather conditions. Surveys should be conducted minimally from the 3rd week in February to the 1st week in April. Sandhill Crane surveys can be terminated for the year in April following a count of less than 5,000 suggesting that most Cranes have moved north. If Sandhill Cranes continue to be present on the river, surveys should go on through week 10 as funding allows. The earliest we have conducted an aerial crane survey was the 42nd day of the year (Feb. 11th) and the latest was the 110th day (April 20th). This equates to roughly the 7th and the 16th weeks of the calendar year as starting and ending survey dates. We attempt to keep surveys as close to 1 week apart as possible beginning on or before 15 February. We often pick a primary survey day of the week and try to stick to it throughout the spring survey season, which has generally been Monday, Tuesday, or Wednesday, considering the high demand for survey results by the weekend.

Table 11. Weekly survey periods

Week 1	Week 2	Week 3	Week 4	Week 5
2/12-2/18	2/19-2/25	2/26-3/4	3/5-3/11	3/12-3/18

Week 6	Week 7	Week 8	Week 9	Week 10
3/19-3/25	3/26-4/1	4/2-4/8	4/9-4/15	4/16-4/22

Surveys begin over the river at about 25-30 minutes before sunrise (the beginning of civil twilight) as soon as Sandhill Cranes in the river are appropriately visible. It is ideal to be leaving the airport in Hastings [or Kearney] at *nautical* twilight, or just before, to ensure surveys are started on time. If light is too low to begin counting at *civil* twilight given sky condition, the pilot can be directed to circle the survey starting point until there is enough light to count cranes accurately. The survey route generally takes between 50 min. and 1 hr. 15 min. to complete based on conditions (e.g., headwinds, number of cranes, etc.). Every effort should be made to keep the survey under 1 hr. and 15 min as Sandhill Cranes often leave the river within an hour of sunrise (Ferguson et al. 1979; Norling et al. 1992).

Survey Route

We normally fly at 700 ft above ground level which avoids disturbing the birds but still allows for accurate identification. We count all birds roosting on the main channel of the Platte River and visible side channels as well as in adjacent off-channel habitats such as wet meadows and corn fields to the distance that we can positively identify and count crane groups from the flight path (Caven et al. 2019). However, we generally only detected crane groups within 3.4 km (2.1 mi) of the flight path and likely at a reduced rate compared to those roosting on the river (Caven et al. 2019, 2020). Furthermore, several groups of Sandhill Cranes will fly beyond this distance from the river to forage during the day, especially as the migration season progresses (Pearse et al. 2015). In this way our survey effort provides an index of abundance that generally represents a significant underestimate of the number of Cranes and is about ~30% lower than the USFWS estimate when surveys are conducted at the same time (Dubovsky et al. 2018; Caven et al. 2020). The flight route totals just over 85 miles from Chapman to Overton, NE (Buckley 2011; Caven et al. 2019, 2020). The survey is divided by 11 bridge segments and surveys are flown from east to west for the first ~7 survey weeks or until peak abundance and from west to east during the last ~3 survey weeks to maximize the total number of cranes detected at riverine roosting sites as abundance tends to peak in the eastern part of the survey area earlier (week 6) than in the western portion of the survey area (week 8; Caven et al. 2019; Table 12). The survey route follows the south channel of the Platte River, which is generally the largest or the "main channel" (Caven et al. 2020).

Table 12. Description of Platte River bridge segments between Chapman and Overton, Nebraska.

Bridge Segment*	Location
1	Chapman to Highway 34
2	Highway 34 to Highway 281
3	Highway 281 to Alda
4	Alda to Wood River
5	Wood River to Shelton
6	Shelton to Gibbon
7	Gibbon to Highway 10
8	Highway 10 to Kearney
9	Kearney to Odessa
10	Odessa to Elm Creek
11	Elm Creek to Overton

* Bridge segments increase from east to west (adopted from Currier et al. 1985).

Surveys are not conducted during mornings of inclement weather (high winds, low visibility, low ceilings (IFR conditions), precipitation, etc.) that could decrease detection probabilities (Ferguson et al. 1979; Buckley 2011; Caven et al. 2020). In the case of poor flight or visibility conditions, the survey should be rescheduled for the following day, if the weather is still not cooperative, attempt to fly the subsequent day. After three attempts it may be reasonable to forgo the survey until the following week considering the budget and the long-term weather forecast. In almost all cases it will be the pilot who cancels the flight due to IFR conditions. It can be helpful to schedule an early morning (~4:30 AM) call with your pilot to check the weather when it is in question before departing from the Crane Trust or home.

Survey Team Roles

Surveys teams include a pilot, an observer who estimates crane numbers, takes pictures for bias estimation, and directs the course of the pilot, and a support staff that records count data, collects GPS locations for each roost, and helps spot for groups of Sandhill Cranes, Whooping Cranes, or other species of interest (Caven et al. 2019, 2020). Over-winged Cessna airplanes generally represent the preferred aircraft that are available for surveys regionally (Model numbers 172, 182, or 185 are all appropriate). Positions of the "observer" (i.e., counter) and the "recorder" in the aircraft can vary per personal preference. However, we generally recommend the observer be placed in the back seat as it provides a longer-duration view of passing roosts. This is especially helpful for very large roosts. The data recorder is then placed in the front passenger seat (right side). As the roost is flown past, the front seated recorder marks the location at the center of the roost as a waypoint and records the count given to her/him by the observer in the back seat. If the roost is continuous and large ($\geq$20,000), the passenger in the front seat will mark the beginning and end of the roost with 2 different waypoints. It is helpful for the data recorder to spot for Whooping Cranes as the observer is counting large Sandhill Crane roosts as well.

Roost Size Estimation

We considered Sandhill Crane groups separated by >100 m as separate roosts following Iverson et al. (1987). Counting large roosts of Sandhill Cranes involves first counting a group of 50 to 100 individuals then creating a mental polygon around that group. That group can then be multiplied in place to account for a small roost (under 2,000) or grouped further into larger mental polygons to count bigger groups (Gregory et al. 2004; Bowman 2014; Drahota 2014; Caven et al. 2019, 2020). Roosts in excess of 20,000 Sandhill Cranes regularly occur and those surpassing 40,000 occasionally appear near the peak of migration (Baasch et al. 2019). In this case a mental polygon can be created around a group of 500 or 1,000 cranes after first estimating the spatial area of a smaller group (e.g., 100). In a sense the same mental polygon technique is applied at two spatial scales in rapid succession to account for these very large roosts (Caven et al. 2019). About 1,000 cranes is probably the upper limit for grouping ("corralling", "bundling") birds accurately. It is important to readjust mental polygons to the density of different roosts across seasons and even throughout one particular survey as the density of roosts can vary greatly (Gregory et al. 2004; Bowman 2014; Caven et al. 2019). Failing to adjust to different roosting densities can greatly increase the bias of abundance indices. Even within a single roost there can be both dense and loose patterns of roosting Sandhill Cranes. Large groups can be circled and recounted when necessary. A second pass is also helpful for taking quality pictures of roosts to verify counts.

Bias Estimation

We assessed the accuracy of our counts by taking photos of a subset of entire roosts along flight path. We took between 1 and 10 photo-subplots of entire roosts depending on the number of roosts detected during the survey, 10 was the maximum number conducted due to time constraints (Caven et al. 2020). We tried to select a variety of roost sizes between 500 and 10,000. We did not assess roosts larger than 10,000 because they were generally too large to photograph in a single frame (Caven et al. 2019). We counted individual cranes in these photos by marking them in Microsoft Paint to produce refined counts for comparison to aerial estimates (Figure 16). When pictures were not sufficiently clear across large roosts to follow this approach, we counted those areas of the roost where individual Sandhill Cranes were visible. We then gridded out the rest of the roost and extrapolated based on roost area to produce a refined estimate to compare with aerial survey data. However, this approach was avoided when at all possible, as counting individual cranes in the photo was more accurate. We generally tried to capture more photo subplots than we ultimately did during surveys as photos had to be quite clear to enable the counting of individual cranes. Artificial intelligence and machine learning may be able to speed up the process of counting cranes individually in photos in the near future (See Akça et al. 2020).

Figure 16. Example of a Sandhill Crane roost counted via photograph for bias estimation using Microsoft Paint. Colors are rotationally used to distinguish and count groupings of 50-100 cranes.

We calculated relative percent bias which considered the directionality of bias estimates and could be used to adjust weekly Sandhill Crane abundance indices up or down (e.g., -15%; Ferguson et al. 1979; Gregory et al. 2004; Caven et al. 2019). It is also important to provide some measure of variability or uncertainty around point estimates of Sandhill Crane abundance. This can be done in several ways depending on the character of the research questions being addressed or the audience being communicating to. Helpful measures that convey variability in bias estimates across roost counts and therefore uncertainty in overall abundance indices include the standard deviation, the standard error, 95% confidence intervals, or estimated absolute percent bias (Altman and Bland 2005; Caven et al. 2019). These are all easy to calculate and supply varying types of information. For instance, the averaged percent bias across all photo subplots regardless of directionality produces an estimate of absolute percent bias (e.g., ±20%). This produces a large confidence interval, that is likely more meaningful on the upper end considering our protocol's tendency to underestimate Sandhill Crane abundance in the region. Using a standard error (SE = $\sigma/\sqrt{n}$) estimate is robust as it accounts for sample variation and size (Altman and Bland 2005). The standard error is intended to measure the level of uncertainty around the sample mean, in this case the level of bias in aerial survey counts. However, the SE can produce a relatively narrow confidence interval that likely underrepresents uncertainty considering our survey method does not account for detection probability. Standard deviations and 95% confidence intervals are also useful methods for communicating variation across survey bias estimates and therefore uncertainty in abundance indices to the public considering the concepts are relatively widespread, if not understood.

Aviation Company and Considerations

Currently, the Crane Trust is flying surveys with sole proprietor Paul S. Dunning (paul_s_dunning@hotmail.com) of Hastings, Nebraska, and departing from Hastings Municipal Airport (40.6143° N, 98.4345° W). However, our provider has changed in the past per availability. Our main contractor was Kearney Aviation (now "Big Air"), but it reduced the number of planes and pilots it retained in 2018 and we began flying with Paul D., who offered more flexibility and experience. Steve Cole (scole@kearneygov.org), Assistant Airport Director at the Kearney Regional Airport, remains a key contact if Paul D. cannot provide aviation services. He will likely have recommendations for available pilots. Big Air also remains an option. They have a single 172 available for rental or charter (Contact: 308-233-5800). The drive to the Kearney Regional Airport takes approximately 35 minutes while the drive to the Hastings Airport requires about 25 minutes from the Crane Trust Headquarters.

Additional Species

Crane Trust biologists also count dark geese (Canada, Cackling, Greater White-fronted, etc.), Bald Eagles, American White Pelicans, Trumpeter Swans, and Whooping Cranes during the survey season. Dark geese and Bald Eagle counts are generally conducted for the first 3 survey weeks of the spring (February into early March) depending on climatic conditions and Sandhill Crane abundance. The goal is only to get a good estimate of the number of dark geese on the river during the peak of their migration and to prepare biologists for large Sandhill Crane numbers. Ultimately dark geese and Bald Eagle abundance estimates do not represent primary objectives and are intended to collect additional useful avian migration data as time permits. American White Pelicans and Trumpeter Swans can be counted throughout the survey season. Spotting these large white birds helps keep biologists focused on spotting Whooping Cranes. Secondly, these species have not been a focus of surveys along the Platte River and this information could prove useful for future conservation efforts. Trumpeter Swans are most abundant at the beginning of the survey period and American White Pelicans are most abundant near the end of surveys in April. Counting Whooping Cranes is a priority of the survey program. The Platte River Recovery Program conducts daily counts; however, they have occasionally missed birds we have detected. Generally, if we can be of help documenting these rare birds it provides additional valuable information regarding Platte River stopover locations. Aside from Whooping Cranes, the counting of species other than Sandhill Cranes represents a secondary priority and serves to collect potentially useful data near the beginning and end of the survey season when a limited number of Sandhill Crane roosts are generally detected. If Whooping Cranes are encountered several quality pictures should be taken of each individual or group, and the detection(s) should be submitted with specific locational information to the USFWS Ecological Services Field Office in Wood River, Nebraska (Current Contact: Matt Rabbe, Matt_Rabbe@fws.gov). It can be helpful to circle Whooping Cranes at a safe distance to garner quality photos of crane groups.

Data Management

Data from each survey is entered into an Excel spreadsheet which is cumulative for each survey year (ex: SACR_20XX_Aerial.xlsx). The spreadsheet includes columns for week, date, observers initials, sky conditions, wind direction, wind speed, bridge segment, waypoint number, picture number, number of Sandhill Cranes (#SACR) per roost, latitude, longitude, the channel where they were detected (Main = M or Other = O), notes (generally indicating survey conditions or non-riverine habitats used), the total Sandhill Crane count for week, and estimated absolute and relative percent error of the survey for the week based on photo-subplot counts. The databases also include columns for the number of dark geese (#CANG+), Trumpeter Swans (#TRUS), Bald Eagles (#BAEA), Whooping Cranes (#WHCR), and American White Pelicans

(#AWPE) detected during each survey. Historically the database also included position within the channel, but we found that the variable was inconsistently applied and therefore we discontinued its collection. If a week is missed due to weather ensure that a line is entered into the database with the week, missed date, and a '*' symbol under the "#SACR" column with additional information in the "Notes" section.

GPS data is vital to nearly all aspects of this study; therefore a few additional steps in data entry can save future researchers significant time. We recommend using the DNRGPS program (Minnesota Department of Natural Resources, Saint Paul, MN) to directly offload waypoint numbers and GPS locations (latitude and longitude) into the database. The data from each morning's survey can be downloaded by selecting the "waypoint" tab in the program and then selecting "download." Choose the file labeled with the appropriate date and the GPS information including the waypoint number, latitude, and longitude will be displayed in a tabular format. This information can be copied and pasted into the Excel database. Additionally, to clarify routes and save time for any future spatial analyses, we recommend saving all roost locations from each survey as a GPS Exchange file (.gpx), which represents a Google Earth, ArcGIS, and Garmin Basecamp compatible file structure. We save each survey week as a new file and keep them in a folder for that year. Waypoints can be quickly uploaded to Google Earth via a GPS Exchange file which can be used to double check the bridge segments associated with each roost.

Ensuring data clarity and accuracy is very important. Any data hand typed should be double checked. We also herein clarify aspects of the Excel database that may not be obvious. In cases where both a "start" and "end" GPS point are taken for a single roost simply enter the Sandhill Crane count, the starting waypoint number, and the corresponding longitude and latitude together in the first row followed by the ending waypoint number and the corresponding longitude and latitude in the row immediately below. Place a "Start" in the notes section of the starting row and an "End" in the notes section of the corresponding ending row. Finally, during data entry ensure that if a bridge segment had no information recorded that you enter a line with '0' SACR for that bridge segment to ensure that future researchers know that this segment was flown and there were no cranes. In the past some bridge segments were flow inconsistently so it can be challenging to tell if the survey recorded 0 cranes or the reach was simply not flown. We do not recommend ever skipping bridge segments. Similarly, any missed data (GPS location, crane numbers, bridge segments, etc.) should be marked with an asterisk in the appropriate column, and explanations or comments kept in the notes column.

References

Akçay, H.G., B. Kabasakal, D. Aksu, N. Demir, M. Öz, and A. Erdoğan. 2020. Automated bird counting with deep learning for regional bird distribution mapping. Animals 10(7):1207.

Altman, D.G., and J.M. Bland. 2005. Standard deviations and standard errors. BMJ 331(7521): 903.

Baasch, D.M., P.D. Farrell, A.J. Caven, K.C. King, J.M. Farnsworth, and C.B. Smith. 2019. Sandhill Crane use of riverine roost sites along the central Platte River in Nebraska, USA. Monographs of the Western North American Naturalist 11(1):1-13.

Bowman, T. D. 2014. Aerial Observer's Guide to North American Waterfowl: Identifying and Counting Birds from the Air. FW6003. US Fish and Wildlife Service, Denver, Colorado, USA.

Buckley, T. J. 2011. Habitat Use and Abundance Patterns of Sandhill Cranes in the Central Platte River Valley, Nebraska, 2003–2010. Thesis. University of Nebraska at Lincoln, Lincoln, Nebraska, USA.

Caven, A.J., E.M. Brinley Buckley, K.C. King, J.D. Wiese, D.M. Baasch, G.D. Wright, M.J. Harner, A.T. Pearse, M. Rabbe, D.M. Varner, B. Krohn, N. Arcilla, K.D. Schroeder, and K.F. Dinan. 2019. Temporospatial shifts in Sandhill Crane staging in the Central Platte River Valley in response to climatic variation and habitat change. Monographs of the Western North American Naturalist 11:33–76.

Caven, A.J., D.M. Varner, J. and J. Drahota. 2020. Sandhill Crane abundance in Nebraska during spring migration: making sense of multiple data points. Transactions of the Nebraska Academy of Sciences and Affiliated Societies 40:6-18.

Davis, C.A. 2003. Habitat use and migration patterns of Sandhill Cranes along the Platte River, 1998-2001. Great Plains Research 13:199-216.

Davis, C.A. 2001. Nocturnal roost site selection and diurnal habitat-use by Sandhill Cranes during spring in central Nebraska. Proceedings of the North American Crane Workshop 8:48-56.

Drahota, Jeff. 2014. Crane Counting During Aerial Surveys 101. US Fish and Wildlife Service. Rainwater Basin Wetland Management District, Nebraska, USA.

Ferguson, E.L., D.S. Gilmer, D.H. Johnson, N. Lyman, and D.S. Benning. 1979. Experimental surveys of Sandhill Cranes in Nebraska. Pages 41–52 in J.C. Lewis, editor, Proceedings of the 1978 Crane Workshop. Colorado State University Printing Service, Ft. Collins, Colorado, USA.

Gregory, R.D., D.W. Gibbons, and P.F. Donald. 2004. Bird census and survey techniques. Pages 17-56 in W.J. Sutherland, I. Newton, and R.E. Green, editors, Bird Ecology and Conservation: A Handbook of Techniques. Oxford University Press, Oxford, United Kingdom.

Iverson, G.P., P.A. Vohs, and T.C. Tacha. 1985. Distribution and abundance of Sandhill Cranes wintering in western Texas. Journal of Wildlife Management 49:250–255.

Norling, B.S., S.H. Anderson, W.A. Hubert. 1992. Temporal patterns of Sandhill Crane roost site use in the Platte River. Proceedings of the North American Crane Workshop 6:106-113.

Pearse, A.T., G.L. Krapu, D.A. Brandt, and G.A. Sargeant. 2015. Timing of spring surveys for midcontinent sandhill cranes. Wildlife Society Bulletin 39(1):87-93.

Appendix 16: Sandhill Crane aerial survey datasheet

SACR Aerial Survey Datasheet (Chapman-1 to Overton-11)

Week: _____ Date: _____ Obs. Initials: _____ Sky: _____ Wind: _____ Start time: _____ End time: _____ Pg _____ of _____

Brdg. Segmt.	Waypoint Number	#SACR	Picture Number	Channel M/O	Notes : (SACR habitat/Behavior, Feb.-BAEA, Dark Geese, Apr. AWPE)	Latitude	Longitude

Notes: "Brdg. Segmt." = Bridge Segment 1 to 11 outlined in protocol. "Channel M/O" = Main channel (M), usually the South channel, and Other channel (O)

Chapter 12: Western Prairie Fringed Orchid Survey Protocol

Project Goals

In 1978, the Western Prairie Fringed orchid (WPFO; *Platanthera praeclara*) was first discovered within the western half of the wet meadows on Mormon Island. In 1982, over 50 WPFO were found flowering in the same area. At maximum, at least 60 plants have been found at the site (Armstrong et al. 2017) previously referred to as "Field 4" which was located in the southwest portion of the current "Northwest Mormon" pasture and another site to the east. This plant species is particularly important because it was protected as a Threatened species under the Endangered Species Act in 1989 (USFWS 1989). Yearly surveys for WPFOs have been conducted since they were originally found, but data reveals a steady decline between 1990 and 2000 (Caven 2022). Despite survey efforts that continued from 2002 to 2004 and from 2010 to present, no WPFOs have been found (Caven 2022). However, WPFOs are notoriously elusive and essentially undetectable in years that they do not flower. The plants are thought to persist in vegetative state for several years until environmental and management conditions are appropriate for the plants to produce a flowering stem. Because of Mormon Island's protected status, the floristic community in the historic WPFO location remains largely intact and comparable to conditions in the early 1980's, which may suggest that the orchids have not been extirpated and may still be found when appropriate flowering conditions are met (Caven 2022). Therefore, yearly WPFO surveys should continue and be a core component of the biological monitoring program. The site at Mormon Island may also be a candidate for reintroduction of WPFO within the Platte River Valley.

Project Methods

Walking transect surveys for WPFO should be conducted within its flowering time window every year. Based on literature, notes, and herbarium specimens, WPFO historically flowered around the first week of July. However, possible phenological shifts in flowering times may have occurred as a result of hydrological changes, management, or climate change. Therefore, surveys should be conducted at least once per week from June 15 to July 15. Transect surveys should cover the historic location of the highest density of WPFO on the west side of "Northwest Mormon" pasture (see blue lines, Figure 17, Table 13). Our surveys will use walking transects modified from Bjugstad and Fortune (1989). Surveyors will walk in parallel lines no more than 30 meters apart as they move systematically in a back-and-forth pattern across the survey area. To cover the entire primary search area more efficiently, multiple surveyors or volunteers may be deployed. A handheld GPS should be used to delineate the corners of the primary search area. Flags may be used to make these corners and to help surveys track where each survey line ends and starts. Effort, including survey duration, the number of surveyors, and the names of the surveyors should be recorded. Survey effort records should be kept in the "WPFO Survey Date Records" Excel file in the "W Prairie Fringed Orchid" folder under the "Vegetation" folder on the X (public)-drive.

In the event of WPFO being found, each plant within the primary survey area should be flagged within 0.5 meters of the plant and a GPS point taken for each plant. A hand drawn map with each of the orchid's locations may also be helpful. Plant community data should be taken for each located WPFO, all other plant species within 1 m^2 should be documented and their covers estimated using a 1 x 1 meter quadrat placed with the orchid at the center. Photographs of each

WPFO and their immediate vegetation community should be taken as well. Located plants should be revisited once per week throughout the rest of the year to monitor phenological progression, recording plant height, number of closed, open, and senescing flowers on each stalk, evidence of seed production, and any incidental pollination visits. The Nebraska Game and Parks Commission – Natural Legacy Program (Gerry Steinauer; gerry.steinauer@nebraska.gov) and the US Fish and Wildlife Service – Ecological Services Field Office (Brooke Stansberry; brooke_stansberry@fws.gov) should be contacted, notifying them of the existence of WPFO on the property. Photography and videography partners like Platte Basin Timelapse should also be notified to assist in documenting WPFO and pollination visits. Flowering WPFOs may indicate appropriate conditions for the plant, and the search for WPFO should be opportunistically expanded to the secondary search locations in appropriate habitat types throughout Mormon Island (see orange lines, Figure 17, Table 13).

Figure 17. Primary (blue) and Secondary (orange) transect lines of the WPFO survey area.

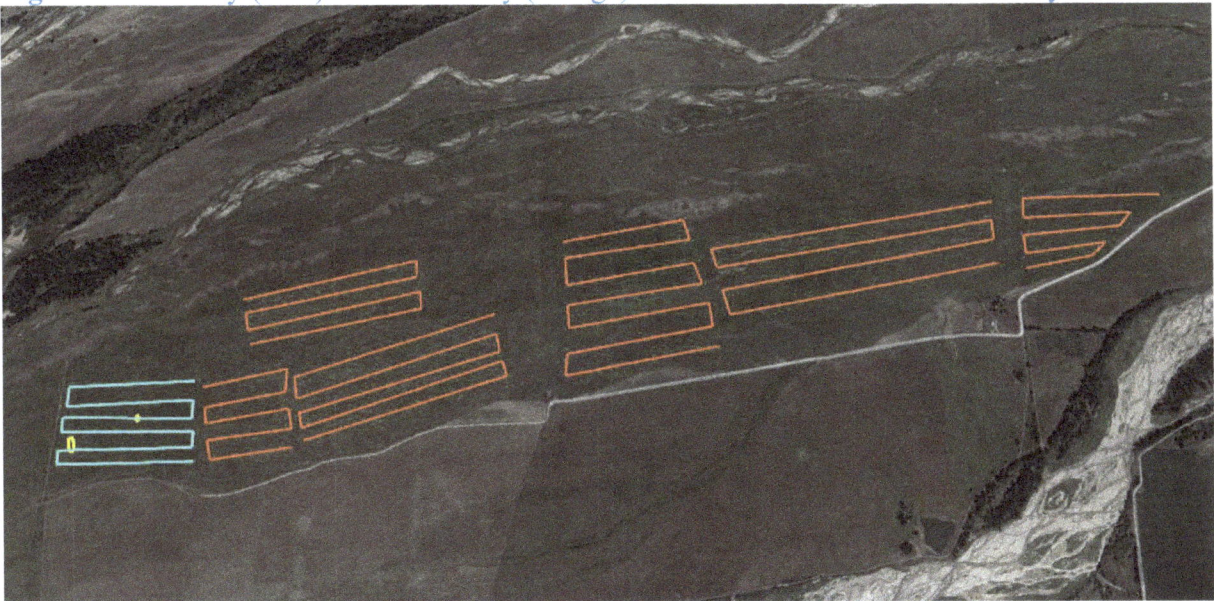

Table 13. GPS coordinates of corners of the Primary and Secondary WPFO survey transect areas

Transect Corner ID	Lat	Lon
Primary WPFO 1A	40.7952654	-98.4302065
Primary WPFO 1B	40.7950358	-98.4354935
Primary WPFO 1C	40.7975671	-98.4356648
Primary WPFO 1D	40.7978806	-98.4306228
Secondary WPFO 1A	40.7958674	-98.4264923
Secondary WPFO 1B	40.7953583	-98.4295624
Secondary WPFO 1C	40.7976493	-98.4302763
Secondary WPFO 1D	40.7983826	-98.4272413
Secondary WPFO 2A	40.7983106	-98.4185041
Secondary WPFO 2B	40.7959143	-98.4261604
Secondary WPFO 2C	40.7979976	-98.4269036
Secondary WPFO 2D	40.8002490	-98.4191843
Secondary WPFO 3A	40.8002707	-98.4222212
Secondary WPFO 3B	40.7989833	-98.4288111
Secondary WPFO 3C	40.8004530	-98.4293294
Secondary WPFO 3D	40.8019840	-98.4225671
Secondary WPFO 4A	40.7993845	-98.4101383
Secondary WPFO 4B	40.7982247	-98.4161004
Secondary WPFO 4C	40.8027640	-98.4164524
Secondary WPFO 4D	40.8036366	-98.4116634
Secondary WPFO 5A	40.8025569	-98.3982132
Secondary WPFO 5B	40.8005754	-98.4095392
Secondary WPFO 5C	40.8026806	-98.4104614
Secondary WPFO 5D	40.8048273	-98.3983029
Secondary WPFO 6A	40.8028285	-98.3945742
Secondary WPFO 6B	40.8024775	-98.3973106
Secondary WPFO 6C	40.8047064	-98.3970807
Secondary WPFO 6D	40.8051679	-98.3911736

Figure 18. Images for western prairie fringed orchid (*Platanthera praeclara*) identification, showing the plant in bloom, a close-up of the leaf shape and structure, and the cauline leaves on a flowering stalk.

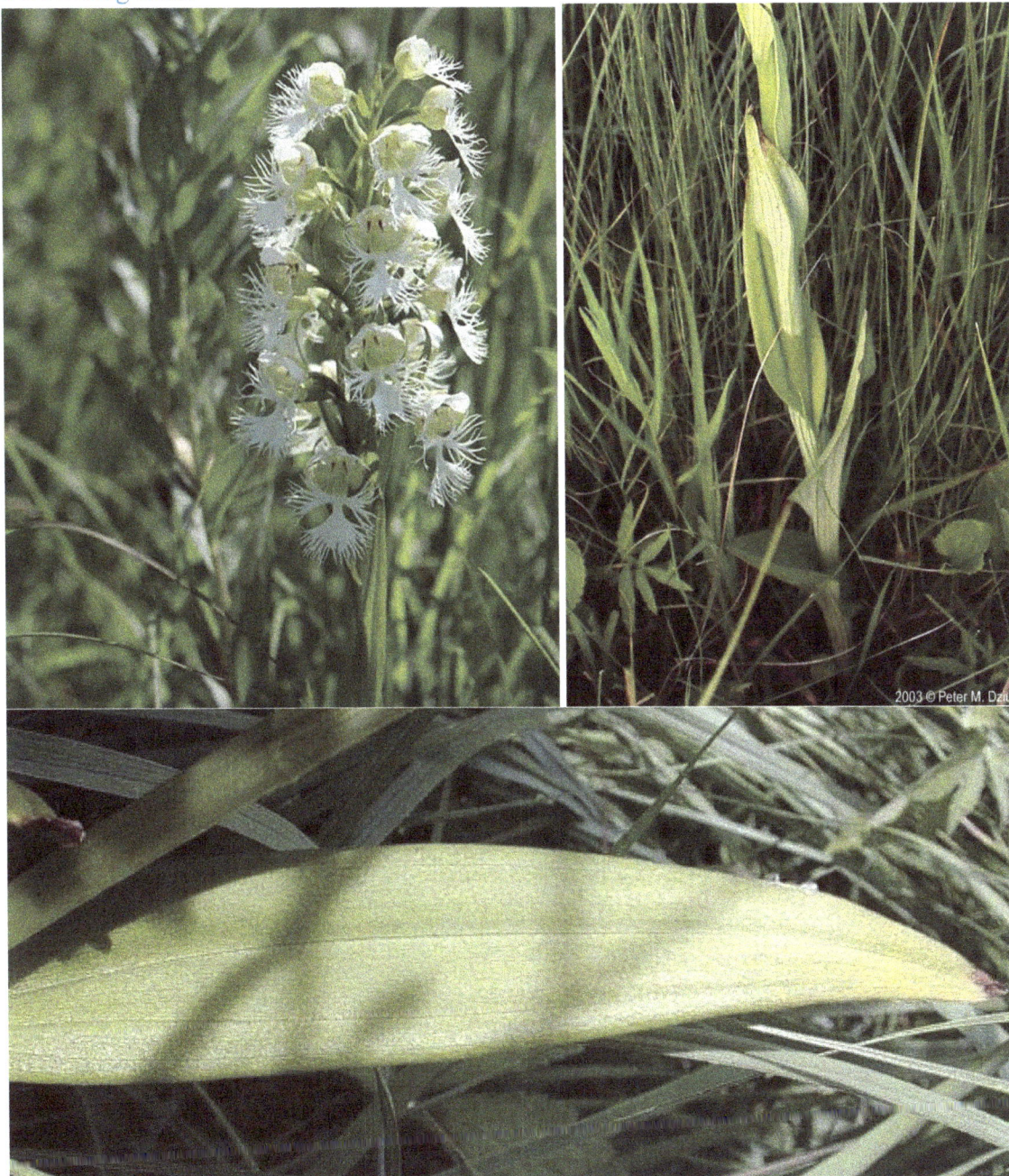

References

Armstrong, D., M. Fritz, P. Miller, and O. Beyers. 1997. Population and habitat viability assessment for the Western Prairie Fringed Orchid at Eugene Mahoney State Park. Final Report, Conservation Breeding Specialist Group, Apple Valley, MN, USA, 128 pp.

Bjugstad, A., and W. Fortune. 1989. The western prairie fringed orchid (*Platanthera praeclara*): monitoring and research. Pages 197-199 *in* T.B. Bragg and J. Stubbendieck, editors,

Prairie pioneers: ecology, history and culture: Proceedings of the Eleventh North American Prairie Conference, 7-11 August 1988, Lincoln, Nebraska. University of Nebraska Printing, Lincoln, Nebraska, USA.

Caven, A.J. 2022. Western prairie fringed orchid management, ecology, and decline at Mormon Island. Transactions of the Nebraska Academy of Sciences 42:1-9.

[USFWS] U.S. Fish and Wildlife Service. 1989. Endangered and Threatened Wildlife and Plants, 50 CFR Part 17, RIN 1018-AB23. Determination of Threatened Status for Eastern and Western Prairie Fringed Orchids. Federal Register 54(187): 39857-39863

Chapter 13: Evaluative Metrics

General Approach

Though we think that having draft evaluative metrics in place is a useful guide, we broadly focus on developing those via the adaptive management process. We have developed some specific thresholds for species of concern that allow us to interpret the availability of particular types of habitats (lowland tallgrass prairie with about 70% litter as groundcover for Regal Fritillaries, Caven et al. 2017; channel widths over 275 m wide for Sandhill Crane roosting, Baasch et al. 2019). However, we often broadly evaluate data using trend lines. For instance, if we have a goal of reducing Smooth Brome (*Bromus inermis*) cover in a particular pasture, we will quantify its total cover via the point-line intercept method per year and examine its abundance trend line over three or more years to determine if targeted management is having the desired effect. Occasionally we will compare plant cover between two years when targeted invasive/exotic species control efforts have been undertaken, but generally analyses will be on longer time cycles (e.g., Caven and Wiese 2022). We also use summary statistics, including interquartile ranges to evaluate biological and environmental variables indicative of management success. For instance, if we find that a particular pasture has breeding avian species diversity under the 25th percentile value, we will reexamine the habitat to make sure we are managing it dynamically. Sometimes lower species richness or diversity is simply a reflection of a habitat's inherent character, however, most of the time we can alter management to make that landscape more dynamic. As we further model our data, we will develop more thresholds on which to base management decisions, but in the meantime, we will be relying on trend analysis targeted at monitoring both species of concern as well as invasive/exotic species. We can also model changes in biological communities over time using trend lines, including species diversity, richness, and abundance across years. The most basic model we use is an ordinary least squares linear regression. For trends that don't fit a linear model we can log transform the dependent variable to fit an exponential curve or include a squared transformation of the independent variable [along with the original predictor variable] in the model to fit a quadratic curve. Generalize linear models (GLMs) are also helpful in many cases. Poisson models work best with count data, while negative binomial models work best with overdispersed count data. Ultimately there are a multitude of analytical techniques that can be applied to adaptive management research at the Crane Trust and thresholds to guide management actions can be directly derived from objectively measurable site conditions thanks to effective monitoring.

Future Directions

Our goal in the coming years is to implement and refine a biological monitoring plan and practical data collection system that allows us to collect the most important and helpful biological data with which to effectively assess the success and variable outcomes of our land management strategies and practices. This will also allow us to study the broader impacts of large-scale biological stressors (drought, flooding, etc.) upon the species we are seeking to conserve. The most important protocols that we will be implementing will be the vegetation and avian monitoring protocols. Their direct link to our conservation mission and overall ecological importance makes them the focal point of our monitoring program. It will be important to collect this information at the ideal frequency with which to detect meaningful ecological change in response to habitat management actions and landscape-level changes resulting from droughts, vegetational succession, and ultimately global climate change. An area of future investigation not mentioned in this text could include simple, local, regular assessments of river channel morphology, to be interpreted in the contexts of the comprehensive data being collected by the USGS (i.e., streamflow). We hope that this document better elucidates the direction of our biological monitoring program and convinces you that biological monitoring is essential to the goals of the Platte River Whooping Crane Maintenance Trust.

References

Baasch, D.M., P.D. Farrell, A.J. Caven, K.C. King, J.M. Farnsworth, and C.B. Smith. 2019. Sandhill Crane use of riverine roost sites along the central Platte River in Nebraska, USA. Monographs of the Western North American Naturalist 11(1):1-13.

Caven, A.J., K.C. King, J.D. Wiese, and E.M. Brinley Buckley. 2017. A descriptive analysis of Regal Fritillary (Speyeria idalia) habitat utilizing biological monitoring data along the big bend of the Platte River, NE. Journal of insect conservation 21(2):183-205.

Caven, A.J., and J.D. Wiese. 2022. Reinventory of the Vascular Plants of Mormon Island Crane Meadows after Forty Years of Restoration, Invasion, and Climate Change. Heliyon 8(6):e09640.

Acknowledgements

We greatly appreciate the Crane Trust researchers that have come before us including Paul J. Currier, Gary R. Lingle, John G. VanDerwalker, Beth Goldowitz, William S. Whitney, Thomas E. Labedz, Craig A. Davis, Robert J. Henszey, Kent Pfeiffer, Felipe Chávez-Ramírez, Daniel H. Kim, Mary J. Harner, Gregory D. Wright, and many more. We would also like to thank the great number of volunteers, interns, fellows, and technicians that have served this research program since 2015 including Katie Leung, Timothy Phillips, Ross McLean, Matthew Conrad, Isabela Vilella, Hannah English, Marin Dettweiler, Carson Schultz, Matthew Schaaf, Amanda Medaries, Alexa Rojas, Brittany Wasas, Charlie Tate, Aurora Fowler, Sam Johnson, Phoebe Dunbar, Sam Heilman, Abe Kanz, Courtnay Pogainis, and several more. Additionally, we want to thank the many partners that have made this work possible including Emma M. Brinley Buckley, the late Robert B. Kaul, Gerry Steinauer, Keith Geluso, Matthew R. Rabbe, Paul S. Dunning, and more. We appreciated editorial reviews of this document provided by Cathy Cargill and Carrie Roberts. Many thanks to Paul Royster, Coordinator of Scholarly Communications for the University of Nebraska-Lincoln Libraries, for helping us publish this document. Finally, we want to thank all those that carry long-term research and monitoring forward for the Crane Trust and in the Central Platte River Valley of Nebraska.

www.ingramcontent.com/pod-product-compliance
Lightning Source LLC
Chambersburg PA
CBHW081419270326
41931CB00015B/3332